OPERATING
FOUR-STROKE
ENGINES

Other titles in this series include:

Operating
Four-Stroke
Engines

Brian Winch

ARGUS BOOKS

Argus Books
Argus House
Boundary Way
Hemel Hempstead
Hertfordshire HP2 7ST
England

First published by Argus Books 1990

© Argus Books 1990

ISBN 1 85486 042 9

Phototypesetting by GCS, Leighton Buzzard
Printed and bound in Great Britain by
William Clowes Ltd, Beccles

Contents

Chapter 1
In the beginning

EVERYTHING HAS to start somewhere and, for the four-stroke engine, we travel back in time to 1862 and we are in the thinking room with Alphonse Beau de Rochas somewhere in Paris, France. In that year, Alphonse published his theory on internal combustion engines, and the factors he set down are those still used in modern four-stroke engines. Compare his concept with what we know are the requirements today for a four-stroke cycle engine:

1 Maximum cylinder volume with a minimum cooling surface.
2 Maximum rapidity of expansion.
3 Maximum ratio of expansion.
4 Maximum pressure of ignited charge.

The requirements for these factors would come from a design function in the following sequence of events:

A Suction during an entire downstroke of the piston.
B Compression during the following upstroke.
C Ignition of the charge at top dead centre and expansion during the next downstroke.
D Forcing out the burned gasses with the next upstroke.

World Aerobatic champion, Hanno Prettner, flies his EZ Laser 230 with a twin carburettor OS Gemini horizontal twin.

7

Sound familiar?

Alphonse never built an engine, but his statement of factors governing economy and performance appears to be the most significant in the development of internal combustion engines. The modern four-stroke engine of today is just as he described in his theoretical concept. Because his work was limited to theory, he was not given the credit for originating the basic principles of the four-stroke engine; the German engineer Nikolaus Otto is the name most remembered for that achievement. In 1876, the partnership of Otto and Langen applied the Beau de Rochas principle to the Otto silent engine. This engine was the forerunner of all four-stroke engines of today and, in 17 years of production, 50,000 engines were sold despite the excess weight, poor economy and 4 horsepower rating. If only they could see what they started...

A rose by any other name

You will hear considerable banter as to the correct nomenclature or name classification of engines, be they two- or four-stroke. There is a cycle of operation for any engine and, for internal combustion engines, the cycle is described as the number of strokes in a single operation. A two-stroke is a two-stroke cycle engine as the piston goes up once and down once for a single operation of induction, power and exhaust. A four-stroke is a four-stroke cycle engine as the piston goes down, up, down and up again for a single operation of induction, compression, power stroke and exhaust. It depends on whether you are a purist or content with the common term when it comes to calling a spade a spade or a non mechanical, human powered, metallic, earth removing tool. I'm quite happy with the adopted, shortened

Horizontal twin cylinder engines provide smooth, low vibration level operation.

name or descriptive of four-stroke or even the friendly name 'stroker' as I know full well what is meant and, by the time you have read this book, I am sure you will be able to join me in this common knowledge.

To begin this story I am going to go through the parts and functions of a four-stroke with you and then we will consider why we choose this complex, engineering marvel over the efficient, simple, powerful two-stroke engine.

The bits . . . inside and out

The head

The head of a four-stroke is where you will see the greatest number of differ-

Quite a few extra bits in a 'stroker' compared to a two-stroke engine.

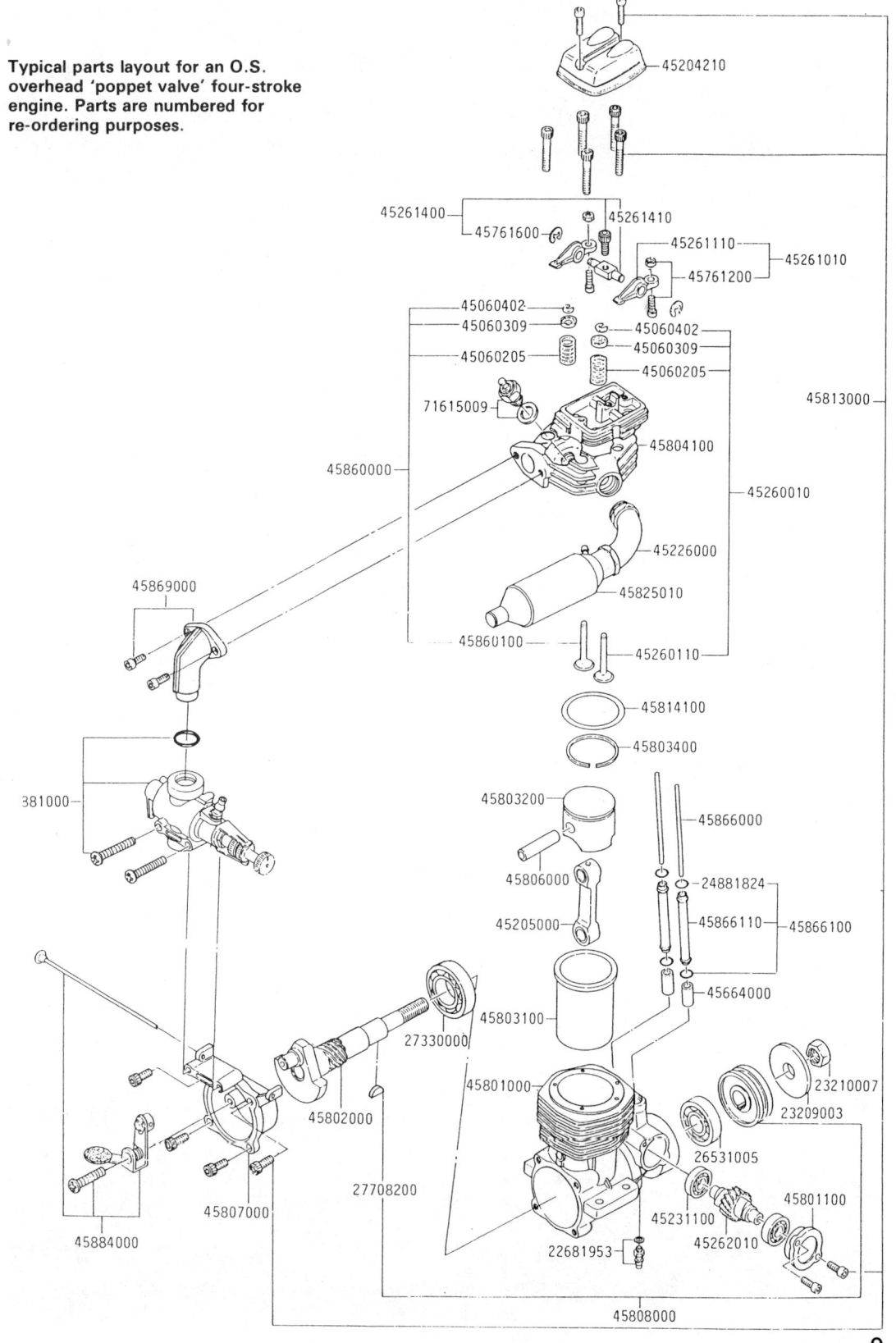

Typical parts layout for an O.S. overhead 'poppet valve' four-stroke engine. Parts are numbered for re-ordering purposes.

45204210

45261400
45761600
45261410
45261110
45761200
45261010

45060402
45060309
45060205
45060402
45060309
45060205

71615009

45804100

45813000

45860000

45260010

45226000
45825010

45869000

45860100
45260110

45814100

45803400

381000

45803200
45866000

45806000

24881824

45205000
45866110
45866100

45803100

45664000

27330000

45802000

45801000

23210007
23209003

26531005

27708200

45807000

45884000

45231100
45801100
45262010

22681953

45808000

9

A 'bathtub'-shaped combustion chamber with one valve removed. The shiny line around the rim of the valve seat is the seal.

ences from a two-stroke. The two-stroke engine has a combustion chamber and ignition in the form of a glow plug or spark plug or, in the case of a diesel, a compression adjusting vernier for the contra piston. (This adjuster is commonly called a Tommy bar.) As we will be discussing an overhead valve engine only at this stage, the first difference you will see in the head will be the shape, extra finning and, in the case of covered rocker gear, a container known as the rocker cover. Under the rocker cover is a little engineering system all on its own. Motion is transmitted vertically to arms that rock horizontally and, in turn, the arms re-convert the motion to vertical force as they, in correct sequence, open valves then release the force to allow springs to close the valves. All of this happens an amazing 5,000 times per minute when the engine is running at 10,000rpm.

So then, what are these parts and how do they operate? On removing the rocker cover we observe two arms that can be merely straight lengths of square

Home-made three cylinder radial with the carburettor set inside the engine mount. Note the double clevice arrangement for the conrods.

metal or forged to the shape of a mandarin or orange segment (with a bit of imagination). These arms are valve rocker arms or, more commonly, tappets (from the tapping noise they make when in action). These tappets are pivoted somewhere near the middle—depending on the mechanical advantage design—on a highly polished shaft that is the rocker shaft. In the rear end of the tappets will be some form of adjustment with the most common being a socket head (for an Allen key) and lock nut and this will be on the top of the rocker. The continuation of the socket headed screw will go through the tappet where it will then be shaped like the head of a slotted screw but, instead of the slot, you will see a half round hole. This is where the pushrod locates. The purpose of this adjuster in both tappets is to set the valve clearance as set down by the manufacturer. This operation will be covered in the maintenance section. On the underside of the other end of the tappets is a convex (outwardly curved) section that is highly polished. It is this

Rocker assembly complete showing rocker arms, shaft and adjusting nuts.

section that contacts the valve stem when pushing it open. It is here that you check the valve clearance with a feeler gauge. The purpose of setting a very small gap between the tappet and the valve is only to ensure that the tappet is not riding the valve and holding it open. If the valve does not completely close the engine may not start, or if it does it will lack power and the valve seat will be burnt.

To answer your next question, the gap increases just a little when the engine reaches correct operating temperature. Only a small section of the valves is visible in the rocker section as they are covered by the valve springs. The purpose of the springs is to hold the valves up tight against the seats inside the head to maintain a tight seal and to return the valves after they have been depressed by the tappets. The springs are secured in position on the valve by a cup and retainer. The cup is more like a small metal dish which locates on, or in, the diameter of the spring and contains a C shape circlip—a retainer—or two split and (generally) tapered collars that

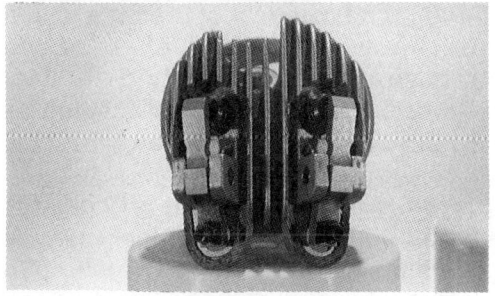

Overhead rocker arms or, by the original name, tappets.

lock onto the valve stem within the cup and these are called collets. Other than the heads of the headbolts and some oil or grease on moving parts, that is all you should find inside the rocker box.

Moving down towards the plug you will see considerable finning on the head and this is there for more than aesthetic appeal. There is a fair mass of metal in the head casting alone, plus all the different bits and pieces in a range of metals which are out of the cooling airstream. If the heat builds up to a dangerous level then something is going to fail or break. The manufacturer has experimented with the finning until an optimum of cooling was provided to keep everything in running order. The reason I tell you this is because it has been known for modellers to 'modify'

Plenty of plugs to choose from so there is no excuse for poor running due to mismatched plug and engine.

The barrel of a four-stroke engine showing inlet and exhaust manifolds. A perfect seal is important to performance.

the finning on an engine to fit in a cowl of a model or in the mistaken belief that more power will be gained if the engine runs hotter. If you 'modify' the fins on the head of your four-stroke engine you are begging for trouble.

We will discuss plug types later on but, for the moment, keep it in mind that the plug screws directly into the aluminium head, i.e. no brass insert as is found in a lot of two-stroke engines. When you look into the combustion area of the head you will also notice that in most engines now the plug is very close to one of the valves. So close in fact that if you should strip the thread there is little chance of fitting an insert or a thread replacement such as Helicoil or the like. The only reason you should have to remove the plug from a four-stroke is because it is burnt out or no longer working efficiently which should, with care, occur only after many hours of running. The removal of the plug to add after run oil is a wasted effort as you only have aluminium and stainless steel in the area, little if any fuel residue as the heat will evaporate it and the ring will be protected by the oil injected into the crankcase. If the engine is flooded or jammed with an hydraulic lock, careful movement of the crankshaft will open one of the valves and the excess fuel can be drained out the opening.

Computer designed crankcase for the English
HI-MAX Magnum engines.

Modern casting and machining is absolute top
quality. A one-piece casting for a Saito crankcase.

Moving down from the plug we encounter the holes for the carburettor tube and the exhaust pipe. These are

manifolds; inlet and exhaust manifolds. Most exhaust pipes screw into the head, as do some induction tubes from the carburettor, and the same warning applies here as I gave you for the plug. Care must be exercised with several operations involving the manifolds. Those that screw in generally use a gland nut—a reasonably thin tube threaded and with a hexagonal end. When tightening these don't lean on the spanner too heavily as they distort quite easily. Before screwing the nut or threaded pipe into the manifold, put a dab of anti-seize type of grease on the threads. If

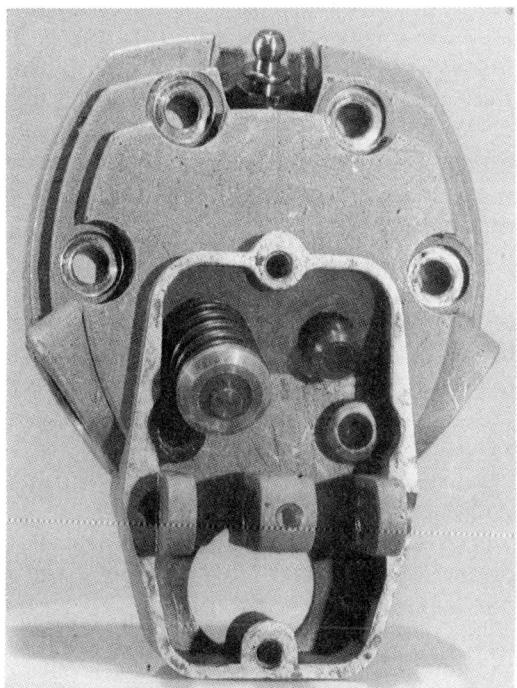

Looking into the rocker chamber we can see the
valve guide pressed into the head.

Saito engines have a one piece head and barrel.

this is not readily available then a little grease or oil will do. This is to ensure the thread screws easily for the full length and that it does not lock in position after some running. Be sure you use the gasket supplied for the manifold as this provides the airtight seal. If the exhaust manifold leaks it could burn a furrow in the metal and may upset the running of the engine not to mention the burnt carbon on that area. A leak in the inlet manifold is a real problem as it upsets the tuning and could cause an engine to cook up due to lean running. The insert type manifolds (where the induction tube screws or fits into the manifold) generally cause little, if any, problems provided the gaskets are well fitted and tightness is maintained. The bolt-on manifold can be a problem if you tighten the bolts too hard. This can cause the manifold to curve and cause a leak. Most engines with this fitting are now fitted with a gasket which does a good job of sealing. If you suspect a problem in this area you can always seal the joint very efficiently with silicone sealant. Both these manifold types have advantages and disadvantages and I would not consider it a necessary point to ponder when making a decision as to which new engine to purchase.

I have had modellers tell me they have problems keeping the exhaust pipes or mufflers screwed up tight on some engines and they asked my advice as to the solution. The first consideration is that if you don't rectify the problem the threads are going to strip out, and that generally means a new head for the engine. There are a number of gasket sealant compounds available for that specific job available from auto garages or service stations. I have two favourites that have been around for so long the tins are now sold with whiskers and they both have different properties. For an (almost) permanent seal I use Stag

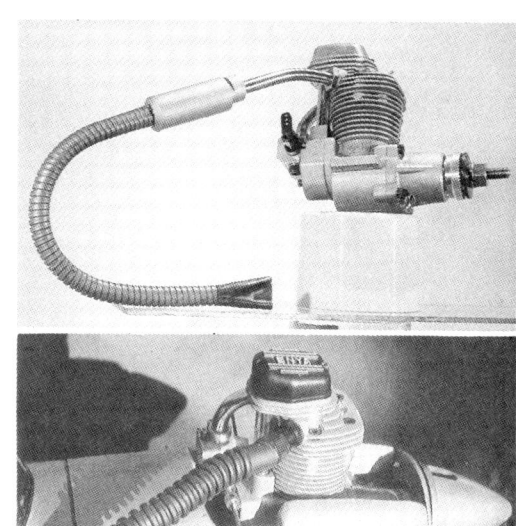

Four-strokes are very tolerant of exhaust extensions as long as you don't reduce the pipe diameter.

jointing compound and for a joint I am sure will have to be undone, but still maintain a perfect seal, I use Heldtite. Both of these are impervious to all the chemicals we are ever going to use in modelling and both will take more heat than we can safely produce with a model engine (disclaimer on both products).

Well, having checked, fiddled, locked or sealed all the components on the outside of the head let us explore the underside and inside where I feel certain we will meet the popular poppet valve. Some mechanics refer to the valves as mushroom shaped but I think you need considerable imagination to come up with that simile. On the underside of the head you will find a depression and, possibly, a thin aluminium or copper gasket. Firstly, handle the gasket carefully as it will crease or bend at the drop of a hot spud. I consider it good practice to anneal this gasket whenever the head is removed. To do this, heat the metal carefully and drop it into cold water to cool. To save a lot of tears and hair pulling, my advice is to heat aluminium gaskets with a burning match unless you are really familiar with propane gas and how easy it is to melt thin aluminium. Copper can be heated with a gas torch but only until you see the green flame. The reason for this is to soften the gasket to provide a perfect seal and this process only applies to gaskets in good condition.

Well, so much for the gasket, let us look into the various shapes of the depressions or, to give them their correct name, combustion chambers. Some years back, if you were in the 'hotty' engine conversations regardless of whether it was full-size cars, motorcycles or model engines, hemispherical was the only way to go. The 'in' terminology was hemi head and a real power machine might be a 'four-valve hemi chamber'. Hemispherical refers to the shape of the combustion chamber or depression in the head and this shape is how the inside of half a tennis ball would look. Nowadays we have new revelations with the 'wedge' or 'pent roof' being the password for performance. If you were going for absolute performance to the nth degree then you would consider this shape quite seriously but, for our purposes, it doesn't appear to make a lot of difference. Some manufacturers go for the pent chamber, some the hemi and one or two have a bathtub shape and they all work quite well. There is perhaps one very small consideration and that is when a modeller carbons up his engine and sends it to me for cleaning—I find the hemi head is the easiest to clean. Describe them how you will as mushrooms, drawing pins and the like but, to me, a valve looks like . . . a valve—what else?

In our engines the valves are stainless steel (all the ones I have checked have been) and are of very high engineering quality. The head of the valve—the round, flat bit—has a tapered under edge and this is the valve sealing angle. It fits tightly to the corresponding angle of the valve seat and is held there by the valve spring we looked at earlier. The rod or long section of the valve is the stem and this is a working fit in the valve guide. When the head is cast, two holes are bored down through the combustion chamber in the position where the valves will be. Bronze inserts are pressed into these two holes to stay (generally) forever. The top section of the inserts is machined down to a spigot of a size to fit the inside of the valve spring. The rest of the work is carried out when the inserts are in place and

I certainly prefer an engine like this to the chainsaw converts. The 50cc capacity swings a 20 inch prop with ease.

A rare configuration for a model engine but a desirable piece of engineering. Powerful and very smooth running. Single throw crankshaft is used in this engine.

15

this includes boring into the side for inlet or exhaust, boring up to form the valve chamber and the valve stem guide, and cutting the bottom edge for the valve seat. Within reason, considering production work and costs, the hole through the side and the chamber are blended into a certain design, and surface finish is quite fine to allow a good flow of gas through the inlet and a rapid exit for the exhaust. Again, caring for this area will be discussed in the maintenance section.

Well, that's about all for the head and I hope you can, now, see that there is considerable difference between a four-stroke and a two-stroke head.

The piston and liner

The first thing you would notice when you compare a two-stroke liner to a four-stroke liner is that the latter is whole in that there are no holes in the side of it. The holes in the side of a two-stroke are ports and these are cut in at specific positions to control the inlet and exhaust timing of the engine. The inlet is also controlled by the rotary port in the shaft or a rear valve in the crankcase. As we have valves for this purpose in the four-stroke, there is no need for holes in the side of the liner or the crankshaft.

This is the crown of a piston. Cutouts are to give clearance for the valves.

These are slipper pistons of minimum proportion to lower reciprocating weight.

You will often find the piston(s) in a four-stroke engine is quite different to a two-stroke piston in that it does not have so much skirt (length below the gudgeon pin) and often has a lot of metal removed from the area below the piston ring. This is to keep the piston as light as possible. The piston is part of what is known as the reciprocating mass, and the lighter this mass is the better acceleration the piston has on the upstroke (less lost energy) and the smoother the engine runs. This cannot be done to the same extent on a two-stroke piston as the length of the skirt and any holes are part of the engine timing. Although you will find circlips securing gudgeon pins in some four-stroke engines, the Teflon or

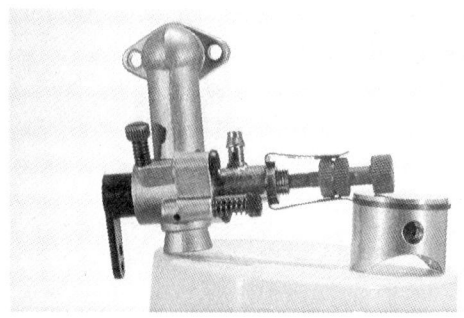

How small can you go? The tiny piston and the seemingly large carburettor is from the OS.20 four-stroke which is the smallest production 'stroker' on the market.

bronze rub pads in the end of the hollow gudgeon are quite adequate as there are no port cutouts in the liner for the gudgeon to catch in as it moves in the piston. The conrod might be just a little stronger in some engines as the forces on this component are a little more in a four-stroke than a two-stroke. My understanding of this is that the forces vary between the power stroke and the induction stroke and a variance of this type would induce an increase in mechanical stress.

The crankshaft

You will see a variety of differences in crankshafts according to the make of the engine. The shaft in a four-stroke performs one extra function over that of a two-stroke; it provides some form of drive for the cams. As an example, O.S. and Saito engines have a gear on the

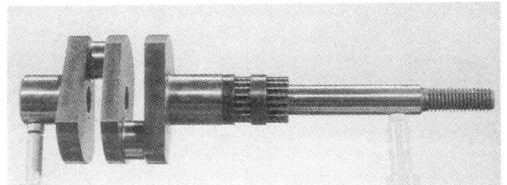

Double throw crankshaft used in horizontal or vertical twins.

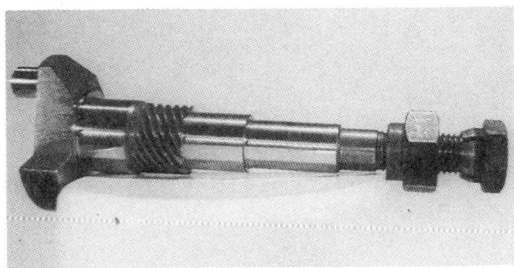

Overhung crankshaft used in single cylinder engines, radial engines and some twin cylinder engines. Small shaft on top left is the crankpin which is coming out of the crankweb or counterweight and from this is the mainshaft which, in this case, includes the gear for driving the cam gear. The correct name for this gear is the pinion.

Camshaft from a British Magnum 120 four-stroke. The shaft is a pressed in roller from a roller bearing.

crankshaft between the main bearings and the purpose of this gear is to provide a 2:1 drive to the camshaft which, obviously, runs at half engine speed. On Enya and Magnum (UK) engines, the crankpin of the crankshaft is extended to drive a secondary shaft—the timing gear shaft—out of the rear of the engine to engage camshafts in a gearbox. Webra use another method whereby a gear belt is driven by the crankshaft to the overhead valve system and H.P. uses 45° bevel gears to drive a vertical shaft to the overhead drum valve used for valve timing. The large diameter of the crankshaft is generally drilled out like a two-stroke shaft but, in the case of a four-stroke, this is simply a weight saving operation.

Additional components

You certainly get more for your money with a four-stroke when you consider all the additional parts of the engine. Around the bottom end you will find one or two short shafts with a gear and some 'lumpy' bits. The lumpy bits are the cams. Sitting on top of the cams will be

Camshaft housing showing gear, cams and cam follower.

some small rods that look like small gudgeon pins with one square end and the other end hollow. In the hollow nests a slender rod with round ends. (Hopefully you will find two of them as one is not quite enough for successful operation.) These rods are called push-rods and they are generally housed in tubes called pushrod covers which are sealed top and bottom with O rings or rubber boots. All these bits work in the following manner: as the crankshaft revolves it turns the gear on it or driven by it at, obviously, crankshaft speed. The gear on the camshaft(s) is twice as large as the crankshaft gear so it turns at half the speed. This is classified as a 2 : 1 gear ratio. If the camshaft gear is turning then the cams must also be turning and here is the reason why they are lumpy. As they turn, the cam followers ride on them and move up and down on the high and low shape. The shape of the cam is defined as: base circle—the round section close to the shaft; the peak—the highest, pointy bit and the sides are the rise—where the follower rides up, and the fall where the follower rides down. The shape (known as the profile) of the cam is an important part of the designed engine performance. An odd point in perform-ance modification is that if you want to increase the lift of the existing cam you machine grind metal from the *base* circle. Can you work out why? (A clue—

Flown from water or, using a 'dolly', from land, the 1930's flying-boat is truly majestic.

Magnificent *Stranraer* flying boat was initially fitted with five cylinder radial engines, later changed to Laser 'V' twin four-strokes.

think about adjustment). Okay, we now have the cam followers moving up and down on the cams so, naturally, the pushrods are also moving up and down and this is a form of transmission of motion or power. Now let us think back a bit to the head. Remember the rocker shafts with the adjustment in one end and the bottom of the adjuster being hollow? Well, that's where the pushrods locate and, as they move up and down, they exert a force to the rockers which causes them to rock as they were designed to do and, as they do, they open the valves.

Back to the beginning

Having finished up back with the valves, we have travelled the full circle and it's time to move on to the next chapter. Before we make the next step there are two final points to consider in our variation between the two types of engines and they are carburettor and exhaust. On most four-strokes the carburettor is attached to the end of some form of a tube. This is the induction tube. Generally it would be better to have the carburettor attached as close to the

head as possible for maximum performance but you can imagine the problems *you* would have trying to locate the fuel tank with the centre line level with, or a fraction below, the line of the needle valve. The difference in performance without the induction tube would hardly be noticeable. A four-stroke engine is quite happy with a much smaller carburettor throat than a two-stroke engine for their designed output, hence we see a (approximately) .25 two-stroke carburettor fitted to a .40 four-stroke.

The exhaust fitted to a four-stroke is generally rather basic being in the form of a bent tube. This is adequate for most engines as the noise level is much lower in pitch than a two-stroke for two reasons. Firstly, the frequency is less due to one exhaust every second stroke and, secondly, the emergence of the exhaust is not as rapid as a two-stroke engine. The small opening offered by the exhaust valve is gradual from closed to open, and is not large enough for the gas to escape in a large load—it is pushed out gradually by the upcoming piston. This action is relative to that of a two-stroke as it actually happens quite fast when you consider 5,000 times per minute at 10,000rpm. With a two-stroke engine the exhaust port is suddenly exposed as the piston goes past, and the emergence of the gas is quite rapid and, consequently, louder. To gauge this effect, put your hand over the exhaust outlet of your car and angle it away gradually. Result? gradual noise. This time remove your hand suddenly and you will get more of a report. (Do this when you first start the car as the pipe will be cool and you won't burn your prop-flicking hand!) The larger four-strokes are a bit quieter with a simple muffler and they are amenable to some exhaust experimenting if you want to drop the noise right down. We will look closer at this later in the book.

Go over all the bits again until you are familiar with each operation and any need for adjustment, maintenance or lubrication.

Chapter 2
Principle of operation

THERE IS no question that a two-stroke engine is infinitely simpler to manufacture, has less moving parts and its principle of operation is less complicated than a four-stroke, but the fact remains that there are far more four-stroke engines used throughout the world than two-strokes. What is the logic behind this? Let us check out the principle of operation of both engines and see if we can come up with some answers.

A two-stroke engine is timed so that the rotary valve in the shaft or rear case is opening when the piston is going up and creating a negative (suction) pressure which sucks the fuel in. The engine fires and the piston travels downward creating a positive pressure which forces the gas (fuel/air mix) in the crankcase to travel up the bypass where it pushes the remaining burnt fuel out the exhaust port and fills the chamber for the next compression stroke. A point to consider here is that the exhaust port is still open when the fresh charge of fuel is entering the cylinder from the bypass port. This port arrangement lowers the charging efficiency in accordance to the rpm of the engine. It is for this reason that full-size engine designers are experimenting with a computer controlled porting arrangement for the two-stroke engine. This sequence of events occurs every com-

The first Enya four-stroke was the 35C with open rockers.

21

Home-made twin with gear driven overhead camshaft.

An original OS.60 FS fitted with a MINIMAG magneto. All the joys of spark ignition without the problems of batteries.

plete revolution of the crankshaft. The position of the ports is governed by the designed performance of the engine which results in a rather narrow power band. To explain this, as an example only, we will consider that ¼ of the engine's power is developed for ¾ of the throttle travel. The remaining ¾ of the power is developed during the last ¼ travel of the throttle. This is the reason you will see a performance type of two-stroke motor cycle burble along in traffic and lift the front wheel when the throttle is opened to pass the vehicle in front. This performance is reasonably acceptable to a sport flier as most flying is done with the engine at full throttle and right in the maximum power band. On the other hand, a pattern flier needs a broader power band for precision flying through the manoeuvres. It is for this reason we are seeing the long-stroke engines (which are more tractable) and four-strokes making their presence felt in the competitions.

The cycle of operation of a four-stroke is an entirely different kettle of fish (ever cooked a fish in a kettle?). Here's how it works: consider the engine has just fired and the piston is on its way down. Just before it reaches the bottom of the stroke—Bottom Dead Centre—the exhaust valve will start to open. There are

two reasons for this and the first is that the piston going down is creating a strong suction that could slow the travel down. Using a hypodermic syringe—the one you flush the engine with—or a pedal cycle pump, push the plunger right down, seal the end with your thumb and try to pull the plunger back up. The resistance you encounter is the

Home-made conversion of a two-stroke to four-stroke engine by fitting new head and belt drive to valves.

same as that which influences the piston depending on the stroke and style of the engine. The slight leak allowed by the opening exhaust valve is enough to relieve the suction. The other reason is that the valve needs to be reasonably well open when the piston reverses direction and starts up again. By the time it reaches the bottom of the stroke the cylinder is full of gas and this would cause resistance to the piston going up if there was not an escape. The piston travels up pushing the spent gases ahead of it and out the exhaust. Just as the piston reaches the top of the stroke the inlet valve starts to open. For a brief moment both valves are open and this is the valve overlap period. The reason for this is similar to that which I explained for the exhaust valve as regards pressure and suction. The exhaust valve is still closing as the piston goes over the top of its travel—Top Dead Centre—and the inlet valve is accelerating its opening. When the cam follower is on top of the peak of the cam, the valve is fully open for a set period. This period is governed by the radius of the peak and this is known as dwell angle or just dwell.

The fact that this description takes very little time to read lulls you into a false sense of time as to how long this valve operation takes. The actual movement of the valves is governed by the valve timing set by the manufacturer and is expressed in degrees or parts of a full circle which is 360°. An example of timing could be: 40° before TDC the inlet valve starts to open—15° after TDC the exhaust valve closes—overlap = 55°. The inlet stays open for 275° total and, if my maths are correct, it closes 55° after BDC. The exhaust stays open for 255°—remember it stays open 15° after TDC so it must commence its opening cycle 60° BDC. Remember that the measurement standard is set by the crankshaft as it

Top end of OPS overhead cam engine.

Close up shows that a rocker lever is still employed to actuate the valves.

travels 360° and it does that 10,000 times per minute at 10,000rpm. If you are still not convinced, watch the rockers on an open rocker four-stroke when the engine is running flat out and see if you can count the 'ups and downs'.

We have established that things are happening fast so let us now return to the inlet valve left suspended while we digressed. As the inlet valve is opening the piston is travelling down and creating the suction to pull the gas (fuel/air mix) in. This continues for the period of time determined by the cam and the timing as set out previously. On the next upward stroke of the piston both valves are closed and the gas is being compressed. At maximum compression and heat, the gas is ignited by the ignition as

23

Constructed originally for the TV *Airline* series, the DC3's Webra 61 two-stroke engines were replaced with Enya 90 four-strokes for a better sound quality.

the piston goes over TDC. The resultant explosion forces the piston down on the power stroke and this re-energizes the inertia of the crankshaft. The piston reaches BDC and... this is where we came in. It sounds rather simple when we talk about it slowly but there are many points of consideration for a successful operation. Firstly, the fuel-air mixture must be correct at 1 part of fuel to 4.5 parts of air for methanol fuel, and 1 part of fuel to 12.5 parts of air for petrol fuel. This is set by the main needle in the carburettor. Next is the load on the crankshaft. If it is too great the engine will not be able to gain speed and will protest in the form of pinging or pre-ignition (explained later). If it is too small a load there will not be enough inertia for the engine to continue spinning over or, if it does, it will over rev or shaft run with disastrous results. Consideration must be given to the adjustment of the tappets or rockers as, if they are too tight or riding, the valves will not close and suitable compression will not be reached to fire the gas. If the tappets are too wide—large gap—the engine will probably run but the power will be down.

The factor that deserves considerable thought is the ignition. With a petrol engine the ignition is by a spark plug and, like the valves, the spark is timed precisely. At starting point at idle the timing would probably be set about 15° BTDC. As the engine gains revs the timing alters (by various means) until it is firing the spark at TDC (depending on the engine design). The simplest method of doing this is by moving the advance lever by hand on a model engine and by manifold vacuum in your family chariot (the thing about 50mm diameter on the side of the distributor is a diaphragm that is activated by the manifold vacuum under the carburettor. It moves in and out according to engine speed and alters the timing by about 15° in the distributor). While this timing is precisely gauged in a petrol engine there is no such luxury in a glow plug engine. We have a modicum of control over the ignition by experimenting with the heat range of the plugs and the number of gaskets on the plug body (the more gaskets, the more the ignition is retarded within reason). If the plug is too hot or protrudes too far into the combustion chamber, the ignition will be advanced and this will cause pinging which is a metallic 'clink' caused by suddenly increasing the load on bearing surfaces. Imagine the crankpin and gudgeon in the conrod surrounded by a cushion of oil. Suddenly, you whack the top of the

What more appropriate than to power a model of a radial equipped prototype with a radial four-stroke engine. This Corsair has a Saito radial engine fitted.

piston with a nylon hammer (dull noise) and drive both the pins to the surface of the conrod bearing and expel the lubrication. You will get a metal to metal ping. This is what pinging is except that, instead of a nylon hammer, the force is derived from the advanced ignition firing before the piston had gone high enough up the stroke. It's a wonder that glow engines run as well as they do with such seemingly random ignition, and it is easily apparent that selection of the correct glow plug is important for optimum performance.

A check list

Having looked reasonably deeply in the principle of operation of both engines let us weigh the pros and cons of both.

1 *Manufacture*. A two-stroke is simpler in construction, fewer moving parts, low cost repairs. A four-stroke is more complex than a two-stroke, more moving parts, but with reasonable care repairs are almost limited to crash damage only.

2 *Cost*. A four-stroke will always cost more than a two-stroke of comparable size but this is offset to some extent by a much longer life.

3 *Efficiency*. Of equal capacity, a two-stroke will put out more peak power but, as a four-stroke develops peak power at lower rpm close to maximum torque, larger propellers can be used.

4 *Noise*. Even when muffled a two-stroke is noisy due to frequency of noise and high mechanical noise. Even at a higher decibel output a four-stroke might not be so objectionable due to lower frequency and more gradual exhaust note.

5 *Fuel*. While both engines will run on the same fuel it is becoming common practice to reduce the oil content in four-stroke fuel due to cooler running and lower rpm. On average, a two-stroke will use twice as much fuel as a four-stroke.

6 *Mufflers*. A four-stroke is very tolerant of muffling and very small mufflers are quite effective. Where a two-stroke would need the muffler outside the cowl of a scale model due to size and the need to keep the large muffler mass cool, a four-stroke is happy with a small muffler or extended pipe inside the cowl with a small exit at the base.

7 *and onwards*. Four-strokes are preferable because they are cleaner running, sound better, look better and are simple, complex little beasties (or big beasties if you go for the large multis that are appearing to gain popularity).

25

By this stage you will have a good understanding of the overhead, poppet valve, glow ignition, four-stroke model engine. When you are ready, move onto Chapter 3 and we will look at some other types of 'strokers'.

Chapter 3
Nothing new but many types

I T IS obvious that you, the reader, has an interest in model engines other than that they are gadgets to mount on the front of an aircraft or in a boat or car to 'make things go round'. As such, I feel certain you would be fascinated to research the history of four-stroke engines and read about the many types of valves employed. I think it would be reasonably close to the mark to say that it has been done, sort of, 'been there, done that', as nothing new in this facet of engine design has come to light for quite a few years. We have had some new designs in engines but the valving has, so far, been either a copy or form of a system that has been in use and proven for so long that you might be very surprised when it was first used.

I will just touch briefly on some common types as a complete coverage would require more than this entire book. Of the most commonly used systems before the popularity of OHV (Over Head Valve), the side valve (SV) was favoured above all particularly for full-size car engines. In a side valve engine, the valve stems are roughly parallel to the piston and open up instead of down. There is a lot to be said for this engine as it was very reliable, low revving and powerful. With the side valve arrangement there is no need for pushrods or rockers and the valve clearance adjustment is incorporated in the cam follower. Due to the odd shaped combustion chamber it was not easy to obtain really high performance (if that is what you wanted) but a noted engineer, H. Ricardo, redesigned the combustion area of the head to give a form of squish to the fuel charge and this improved matters considerably. As OHV was the 'in thing', little further research was done on the SV and it is now most only seen on stationary engines or rural machinery. Pity really. Have a look at the drawing and imagine a contra piston in the head for a side valve, adjustable compression, model diesel engine. Hmmmm, must look into that. Before we leave poppet valves, it is worth mention-

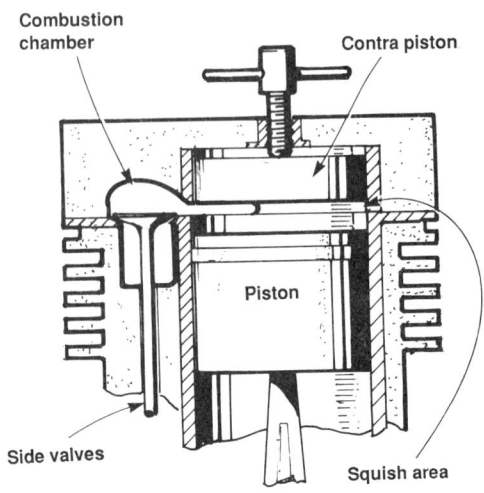

'Dieselized' Ricardo head.

Combustion chamber

Contra piston

Piston

Side valves

Squish area

ing the 'got to be different' engine made by Rolls-Royce and Rover for a number of years that proved its reliability in the many Land Rovers that travelled millions of miles on epic and exciting journeys through wild countries and vast plains. The engine was a compromise of both systems in that it had an overhead inlet valve and a side exhaust valve. It was finally considered to be of little advantage over the conventional OHV so it was relegated to the dusty shelves of engineering history. Again it may be worth another look these days with the constant search for ways to obtain more valve space. Can you imagine, in this instance, that type of engine with an inlet valve the same diameter as the piston? At least it would be the ultimate in one-upmanship over the four valve heads.

There are a few more variations on, not so much the valves, but the methods of driving them. The two most common methods in the unusual or high performance type engines are overhead cam (OHC) and desmodromic. With overhead cam the camshaft is really

overhead, i.e. on top of the head rather than in the crankcase or a gearbox attached to it. Of course it still has to be driven at 2 : 1 speed and this is usually by a chain or, more commonly these days, by a toothed rubber belt. The advantage of the OHC is the elimination of all the ancillary equipment such as cam followers, pushrods and rockers which all take a little power to drive and also have limitations to their efficiency, particularly at high speed. The better known OHC system is where the cams actually contact and actuate the valves but there are variations on this. The 120 four-stroke made by O.P.S. is an example of a variation as they have the overhead camshaft driven by the tooth belt but rockers are used to transmit the forces to the valves. One of the reasons for this setup is that, through mechanical advantage, the drive of the cam can be magnified by having the fulcrum or bearing point of the rocker closer to the cam than in centre. You employ this same law of leverage when you connect the pushrod in your model to the last hole out on the servo output arm and the closest hole in on your (example) stabiliser. You are magnifying the movement of the servo for maximum deflection. Get the picture?

Desmodromic is the name of a valve drive system where the valve is actuated upon for all its travel by the cam. Simply, levers and rods are connected by clevises through the entire drive chain and the valve is driven open and pulled shut therefore dispensing with the need for strong valve springs and the rods and followers. Very efficient, expensive to manufacture and a bit complex for the small size of our engines. No doubt some model engineer has considered the idea or, maybe, made a system for an engine but I don't think you will see too many on the flying field. Next time you see a Ducati motorcycle with

'Desmo' attached to the name you will now know what it means and that it is not the name of the rider!

There are many other forms of valving that I will leave for you to read in other books as I don't have space here. Look for sleeve valves—the liner turns, piston valve—poppet valve in the piston or form of two/four-stroke and many others. Some are really complex and of no great value other than to indicate that someone has been keeping up one of our strange traits of 'doing it a different way'. Some of the systems have resurfaced for consideration lately due to a greater knowledge of metallurgy and lubrication as it was generally the lack of this knowledge that caused them to be relegated to the 'look at later' shelf. One of these is the Aspin head which is used on the Webra T4 and a similar system is also used on the British Robinson RVE engine. To visualize the Aspin principle think of an umbrella blown inside out and one panel missing. The 'handle' of the 'umbrella' goes through the top of the head where it is fitted with a bevel gear. The cone shaped part with the hole is a lapped fit to what would normally be the combustion chamber of the head. This combustion chamber area has three holes (plus the bearing hole for the centre shaft) in order of inlet, glow plug and exhaust. In operation, a sprocket on the crankshaft drives a tooth belt at a 1 : 1 ratio to a shaft in the head with a 15 tooth bevel gear. This gear engages the 30 tooth (our 2 : 1 ratio) bevel gear attached to the 'busted umbrella'. As all this turns, the inside cone revolves, opening in turn the inlet (fuel charge in), the glow plug (ignition) and then the exhaust (gases out). Of course, while this is going on the piston is moving up and down at the appropriate times and this drives the crankshaft which drives the belt that... well, we've been there,

Hirtenberger Patronnen went one different with their 'stroker' by fitting a unique rotary valve in the shape of a drum valve.

done that. The principle of the RVE is the same but it has a flat disc with a shaped section removed and the drive is by gears and a shaft.

The HP (Hirtenberger Patronen-Austria) range of four-strokes is rather unique in their valve layout. The designer took a gamble, defied a few stated principles and the gamble paid off. Consider a rear drum valve in a two-stroke engine, put a gear on the closed end and fit it vertically in the head of an engine and you have an HP four-stroke. Unfortunately, the HP range was introduced at the same time as some innovations that were popular in model engine manufacture and they, like several other makes of engines, had a few minor mechanical problems. Modellers, being rather perverse creatures, are a very unforgiving lot and nowhere is the adage 'give a dog a bad name' practised more vehemently than on the flying field. Even though HP repaired, at no charge, any engines that had the problem, some mud still sticks and this is a shame as the engines are unique, reliable and, seemingly, run for ever. Of the letters I have received from contented owners the general consensus is that you can't stop them (meaning

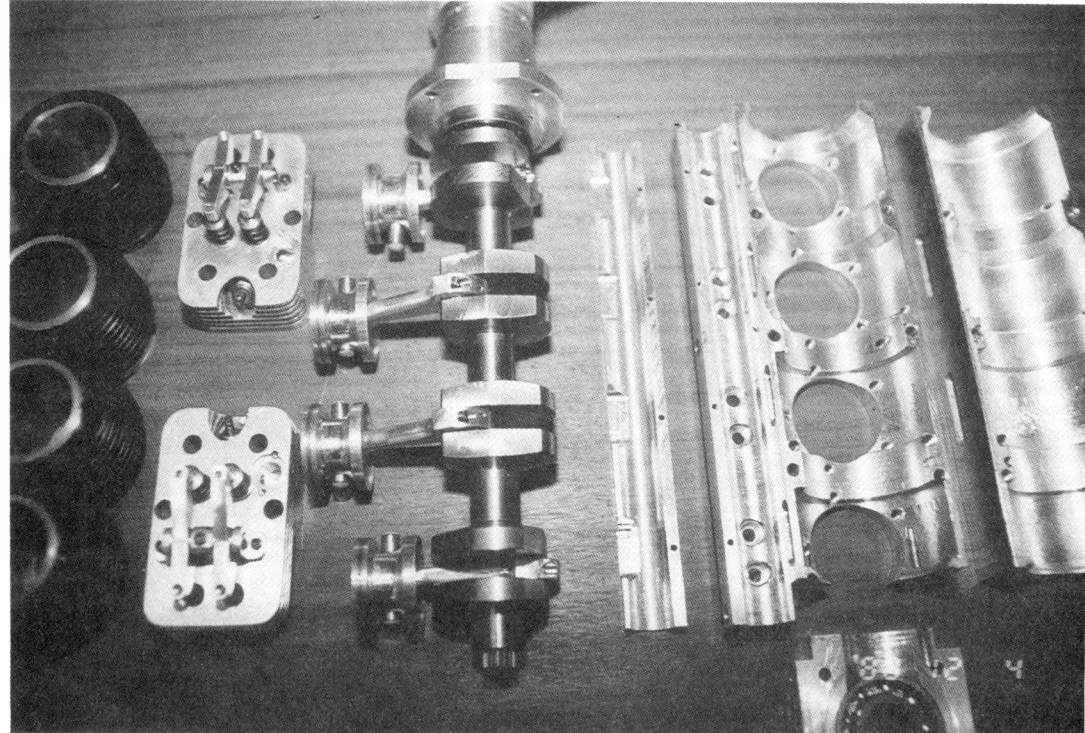

Lloyd Ressler decided to design and produce his own four cylinder in-line engine for his scale *Tiger Moth*. The 50cc engine uses Laser cylinders, pistons and modified heads, and is highly successful.

they keep on running regardless of age) and the exhaust note is like the sound of the buzz of a sewing machine. One other point that some modellers don't like is the front mounted carby but, if you look at it objectively, it is in the same position as most two-strokes except that it is upside down. If you are still not convinced, make a simple manifold and move it around to the side of the engine under the plug. I have done this for a customer and it made no difference to the running qualities of the engine.

Overhead rotary valve

Considered a newcomer to the field, this type of valving actually dates well back in engine history. I have detail of it in my gasoline engine encyclopedia which is the twentieth printing in 1943—the first edition being 1910. This system would have also been one of those relegated to the 'try later' shelf due, in my view, to limitations of metallurgy at the time. The modern day version is a single valve or tube in the head and labours under the name of horizontal axial flow rotary valve and it is this system that is used by the Webra four-strokes. There is a lot of undeveloped potential in this system

The 'how many stroke' Wankel from OS and Graupner. This is the latest version with a capacity of .30 and a performance to match most hot. 45's.

and it is much more economical to produce than a poppet valve system. I am surprised that the twin rotary valve has not yet appeared on the model scene as it is even simpler than the single tube, has greater potential for performance and is the absolute answer for a diesel four-stroke. The rotary valves would need to be driven at ¼ engine speed—4:1 ratio—but this is no problem with the quality of tooth belts available now.

That's a reasonable grounding on valve types for a start so we will leave it there while we have a brief look at layouts of engine types.

Chapter 4
Configuration of engines

SURELY YOU all have seen the various multi-cylinder engines now available—or at least a picture in a magazine—so I won't wax lyrical about them in full. The main reason I am writing this section is due to the number of readers of my magazine articles who write to me asking how the crankshaft looks or works in a such-and-such engine. So, let's do a brief roundup of the various types. The most common type used in single cylinder engines is the overhung crank. Very simply, the crankpin (the little bit that fits into the big end of the conrod) is unsupported on one end. If there is another web and shaft on the end of the crankpin it is then a full crank. The overhung crank is used, as mentioned, in single cylinder engines, vee and horizontal twins and, surprisingly, radial engines. A vee or horizontal twin can be set up with inline or staggered cylinders. If the cylinders are in line, the big ends of the conrod are generally clevised. To do this, one big end is wide with a slot cut

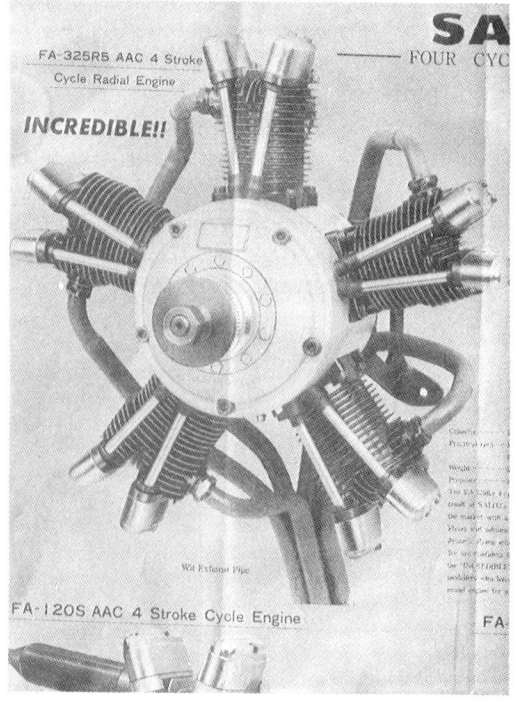

A desirable engine but this is probably as close as a lot of us will get to it (a pin-up on the workshop wall).

32

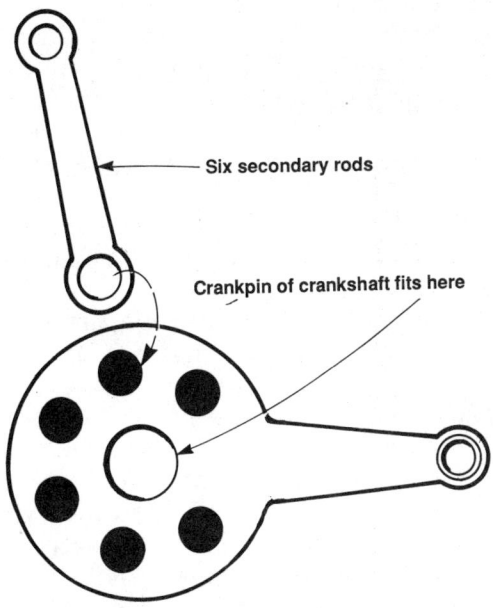

Six secondary rods

Crankpin of crankshaft fits here

Master rod for radial engine.

The OS Gemini range of twin opposed engines are popular for scale aircraft due to the real appearance and plenty of pulling power.

Everything is tucked neatly under the engine which makes it very easy to fit in a cowl.

up the middle in line with the shank and the other big end is narrow and fits in the slot of the other. This way the shanks of the conrods are in line on the one crankpin. If the cylinders are staggered, a longer crankpin is fitted or machined on the crankshaft to allow

For multi-engined models the sound of four-stroke engines running in unison is a modeller's delight.

O.S. have now developed a supercharged version of their 120 Surpass engine to give more power for the competition aerobatic enthusiasts.

a normal rod at the top or little end, but the big end comprises two large discs with a hole in the centre to fit the crankpin and holes around the outer diameter to receive the secondary crankpins for the slave rods. To answer another common question while we are talking about radial engines—why do they have an odd number of cylinders? Mainly for timing purposes. With a firing order of (simply) fire-miss one, fire-miss one etc, you can see that it would not work if you had an even number of cylinders (try it on paper). If the cylinders fired one after the other the oscillation of the engine would tear it off its mount or it would destroy itself. The other type of crankshaft used for some twin and all multi-cylinder engines is a multi-throw crankshaft as used in OS Gemini, Pegasus and the larger Saito twins. These shafts have, at least, a bearing

two big ends side by side to be fitted. In a radial engine it is the conrod that is complex—the crankshaft is just a normal, overhung crank. The 'master' rod is

Enya 53 four-stroke is neatly hidden away under the dummy *Hermes* engine in this DH60 scale model.

support at each end and, in the case of some multis such as your car, a number of main bearings. You will generally hear this referred to as 'a 3 (or whatever) bearing crankshaft' and this means the number of inline main bearings.

There are other configurations such as two crankshafts, epi cyclic cranks and a recent redevelopment with cylinders like those in a revolver (firearm) but not of great modelling interest so how about we move on to setting up your engine?

Chapter 5
Setting up and running your engine

WELL NOW ... happy day! You've just purchased a shiny new four-stroker and you're itching to get it running. Hold on a bit while we check this beastie over; after all, you wouldn't jump into a new car and drive off without checking oil, water, tyres and fuel would you? Remember that we must always consider the human element and we are all prone to make a few mistakes. If something is not set up correctly or *we* don't follow the correct procedures, nasty things can happen. With modern engines there is little chance that something is not correct but ... there remains a remote chance that something is *not* correct. Check the engine over visually to see if all looks okay. Try the bolts for tightness. Take the plug out and turn the engine over several times; sound okay? That little clicking noise you hear is the cam followers and nothing to worry about. The engine should turn freely without binding and no catching in the bearings. Remove the tappet cover and check the gap. Are the tappet adjusters locked tight? Small amount of grease or oil on moving parts? Which way does the carby work? Is forward on the arm open or closed throttle? Stop screw set for idle? Is it an air bleed or needle adjustment for idle mixture (necessary to know for priming)? It might be set out in the instructions that all settings are preset to a nominal running position at the factory but, is the engine now still correctly set?

If the model shop owner had it on display for a while, consider how many modellers might have been in, looked at the engine and even handled it. You know what modellers are—fiddlers! Knobs would have been turned, settings upset and nothing returned to the original setting. Having the same desire as you for the engine but not the ready cash, the 'looker' would have returned the engine to the shop owner who would replace it in the showcase ready for the next 'looker' or, hopefully,

Another successful flight with Dr Jeremy Shaw's *Stranraer* proves the reliability of the Laser 'V' twin engines—they never miss a beat.

purchaser. You dropped in, in a display of pure solvency threw the cash casually across the counter and walked out with the engine, complete with odd settings, under your arm displaying to all the world a look of confidence in your knowledge of matters four-stroke. (After all, you have read this book.) You could bet a fair wager that the internal parts of the engine have been well oiled on assembly but . . . not all bets are winners! With a syringe or a small funnel and some fuel tubing, load about 10 to 15ml of the fuel you are going to use into the engine via the crankcase breather and shake the engine around a bit. It will all blow out again when the engine is started but, even if your bet was a winner or a loser, you still win as the engine will not be starting dry. Put a tiny dab of grease or machine oil on the threads of the plug and screw it into the engine snugly without straining every muscle in your arm. I use a tube spanner with a 50mm (2 inch) cross bar and tighten only with my fingertips. Nope, never had a plug take off on me!

Now that you have cast a critical eyeball over the engine and made sure it is ready to run, how about reading the instructions? Read what the manufacturer has to say about his product— fuel to use, propeller sizes, running in procedures. It helps to give the engine a little bench time if, for no other reason, just to familiarise yourself with the adjustments and starting procedure. This 'bench running' can certainly be in a model but still run it on a 'bench' as opposed to running it on the ground. No matter what the surface, grass, concrete or tarmac, it will always have abrasive dust on it and your propeller will suck it up and the carburettor will suck it down with predictable results. My test bench is waist high and I never completely strip it of a light oily coating as this is ideal for trapping airborne dust and gluing it down.

Right then, having read the instructions, the engine is now locked in your engine test stand (a great investment for *all* engine modellers) or firmly bolted to a timber mount which is tightly clamped

Beautiful DH89a Dragon Rapide is powered by two O.S. four-stroke engines, but will maintain height on one engine. Fully cowled engine sounds great.

to the bench. (You wouldn't even consider clamping the engine in a vice would you? Just the thought of it brings me out in a cold sweat.) Select a balanced propeller in the size range set out in the instructions that will allow the engine to run lightly loaded (not too big on diameter or too coarse on pitch) and (preferably) one with a bit of weight such as a fibre reinforced plastic or fibreglass. If you only have a light wood propeller, a spinner or spinner nut will help as you need a bit of flywheel action when the engine is new. I have a balanced, brass flywheel about 55mm diameter and 6mm thick that I fit with the prop for first starts on some engines if they are inclined to be a little tight or have a tendency to kick back when starting. We established that the propeller was balanced before fitting, now lock it on *tight*, remembering that a shaft run for a four-stroke would almost always be fatal—for the engine that is. (A shaft run is when the engine runs without load such as kicking the propeller loose and then running with no weight other than the crankshaft. They scream like crazy... for a few seconds, then go awfully quiet until you attempt to turn them over and you hear noises that you shouldn't.) Right then, what about the tracking of the prop? A propeller that doesn't track correctly is just as bad as an out of balance prop and it can shake that new whizzbang you just spent all your yearly allowance on to a pile of loose fitting, useless bits. Make the effort—it only takes a little moment of the amount of enjoyment time you will have if you look after your investment. All you need is a bit of Plasticine or Blu-

Tack (the re-usable sticky blob for many a sticky job—also great for holding screws on the end of the screwdriver for tight spots), a sharp pencil or a short length of stiff wire for a pointer. Attach your 'blob' to the bench in line with the tip of the prop and stick the pointer to it so that the pointed end is just touching the back of the prop right at the tip. Rotate the prop carefully and check the distance between the pointer and the other prop blade tip. It should be the same as the first. If it is more than 2 or 3mm out you will have to sand the back of the prop hub to correct it. If it is a long way out, fill the hole in the hub with epoxy and re-drill it or use it as a paint stirrer.

Fuels

Righto then, all set up and (almost) ready to run. Just one little problem: fuel! Did I say a *little* problem? Wait until you ask some of your modelling mates for advice. You will get as many varied answers as you would find feathers on a duck. *Everybody* has *the* best formula. Well, I'm going to tell you now, a model engine *will* run on a great variety of fuels. All you have to do is select one that is correct for you, your engine and for the protection of the parts of your engine. The manufacturer will list a fuel mix in the instructions that will give maximum protection for the leanest run and the engine will run on this mix quite well if not a little (read 'a lot') oily at the exhaust end. The general recommendation will be 20% castor but I have noticed a few 15 to 20% castor or synthetic recommendations coming up in new engine instructions. What do we go for?

Before you jump in, consider the purpose of the oil in the fuel. Okay, it lubricates: what does that mean? The function of lubrication is to prevent metal to metal contact, provide a slick surface for moving contacts and a cushion for tolerance impact. So? A 'U Beaut' super thin synthetic will be ideal as it can wick into the small spaces and do all the things mentioned. Not quite! Consider first that the lubrication for our engine first must undergo combustion pressure and heat as it is mixed in the fuel. A little bit squeezes past the piston and gets down to the bottom end. If the oil can't withstand the first onslaught or ordeal by fire it is of little further use.

Check points

Point one is that the lubrication must be able to withstand temperatures in the range of 190 to 260° Celsius (375 to 500° Fahrenheit). The running temperature of our engines is, ideally, around 190°C but some of you needle twisters can get up to (and above) 260°C on a lean run and, with a melting point of 660°C when aluminium becomes a liquid, you can see that a very lean run could give the top of the piston a case of the jelly shake formation. On many occasions I have seen the tops of pistons of two-stroke engines go granular from a close encounter with a melt down situation, but I doubt that this would occur with current four-strokes due to the cooling stroke (non-firing). What would occur would be a complete breakdown of the lubrication followed closely by a mechanical failure of some type.

Point two is the surface adhesion of the oil. It must have a property or capability to cling to all manner, shape and position of surfaces in a fine film as opposed to globules. Some man-made (or modified) lubricants contain a surfactant to give the oil this property. One example of a surfactant is detergent.

Pour a spoon of water on a sheet of clean glass and it will roll around in globules and not cling to any one area. Add a drop of detergent and the water will become a film on the glass that can only be removed by vigorous rubbing with a very dry cloth. Engine oil needs this property and it is very apparent in one of the oils we use, much to the chagrin of antiseptic modellers who must have spotlessly clean surfaces on their model.

Point three is the ability to carry away harmful material such as burnt deposits, metal particles, ingested foreign material and chemical residues. To be able to do this job, the oil needs to have certain properties whereby it maintains a reasonable viscosity, does not produce too many foreign factors itself and, upon expulsion from the exhaust or sump relief, it will carry away any of these foreign nasties. Again, one oil stands out rather well in this department. (I digress here a moment to explain viscosity. Simply, it is a measurement of the thickness of the oil. Again, simply, a specified amount of the oil being graded with this viscosity is poured through a standard funnel and is carefully timed. The time it takes to run through is then referred to a standard and the viscosity can be read off the scale. The most common scale we use is SAE—Society of Automotive Engineers. This is a very simplified explanation of a very technical process but it gives you the correct idea.)

Point four is one that is not known or understood by a lot of modellers and it is one that is most important—the ability of the oil to carry away engine heat. It is an old saying among mechanics that if you are using a drop of oil in your (car) engine, at least you can be assured that it is going through. Perhaps only a conscience-easer when your chariot blows smoke but certainly true for model engines where we use a total loss

oil system. If you are stupid enough to hold a finger close to the outlet of the engine exhaust, you will find that it becomes very oily and burnt. The burn comes from the heat carried by the oil and this carrying of heat is an important part of engine cooling. Again, one oil stands alone in this category.

The final point is the ability of the oil to withstand the hammering of moving parts and, believe me, there is a lot of this going on in your engine particularly if it is a four-stroke. To illustrate one form of hammering, we are going to play a little game. With your left index finger and thumb, form a circle. Into the centre of the circle I want you to place your right index finger. Now, move your left hand up and down causing your right index finger to strike the top and bottom of the circle alternatively. Any noise? I didn't think so, but imagine if the parts of the game were metal and the movement occurred 10,000 times per minute. To some extent, this type of hammering is going on in your engine at the big and little end of the conrod and at the bosses of the piston where the gudgeon pin is fitted. Yes, I know, a rotary action takes care of this but, not all the time! When the bearing areas are manufactured they have to have a tolerance—a 25mm pin won't fit a 25mm hole without a hammer. Either the hole is bigger or the pin is smaller so there is a minute gap for a running fit which increases in proportion to the use of the engine. Change of speed, increased load, engine cough or backfire causes the bearing to hammer. (Ever heard a tired car engine chattering up a hill? Big end rattle!) Due to direction of motion the piston bears harder against one wall of the liner (cylinder) on the upstroke and vice versa. The position of the crankpin and angle of the conrod creates this uneven pressure and the change occurs at the top and bottom of each stroke. As the piston wears, the hammering contact increases until it is quite audible and this is the traditional sound known to riders of two-stroke motorcycles when the speedo is showing some considerable distance. It is called piston slap.

In a four-stroke engine there are other hammering areas and, again, we are going to play a 'handies' game. Spread the fingers of your left hand to three V's. Do the same with your right hand and then place your right little fingertip against the outside middle of your left little finger at about 90°. Now, rotate your right hand in an arc over your left hand, letting each right finger contact the tip of each left finger before dropping into the V. This demonstrates the meshing of a spur gear and the hammering effect the teeth have on each other. With the number of teeth on the largest gear in your engine you can calculate how many times this happens every minute at 10,000rpm. There is also a hammering action occurring when the cam followers move on the cams. The spring on the valve stem transfers pressure across the rocker arms (tappets), down the pushrods and onto the cam followers. As the cam profile changes during rotation there are changes in speed of the cam follower—acceleration, deceleration— and, according to the profile of the cam, a certain amount of hammering can occur. (This is the clicky rattle you hear that raises the hair on the back of your neck when your first turn over your brand new four-stroker.) Before we get paranoid, all these hammering effects are minimal—if they were a visible object you would need a microscope to see them—but, they are happening and I liken them to the ancient Chinese homily 'dripping water wears the stone'. Put a sponge under the drip and the wear stops. Use the correct oil and the

metal to metal contact stops at the hammering points. You might consider I am labouring the point a bit regarding oil (I'm not finished yet!), but you are reading this book to help you understand your hobby.

As I mentioned at the start, almost everybody on the field has a special fuel brew but what knowledge lies behind that formula? A lot of modellers base their ideas on full-size engines and this is where we can come to grief. We consider an oil that might suit our purposes because of its intended use in the full-size engine. Right then, let's consider a few points. It is intended to be used diluted with fuel at a dilution rate of 4 : 1 or greater? Will it work or mix with methyl alcohol? Will it still work after it has been subjected to combustion heat and stresses? Will its character change if we add aromatics such as nitro methane? Will it withstand the rpm our engines run at? What protection will it give if the engine temperature rises dramatically during a lean run? What sort of protection does it offer when the engine is set aside and also when it is re-started? Are you prepared to test it in your 'stroker'? You will sift through all these 'ifs' and some of you will come up

Neil Tidey's English made Laser engines are machined from stock material. Powerful without the need for nitro.

with a synthetic that meets all the criteria and is used in outboard engines (petroil lubrication), go-carts, surf and snow skis (powered) many of which also use a petroil mixture (petrol/oil mixed) for lubrication. One small problem! Those engines are going to be running a ball or roller bearing in the big end of the conrod and some also in the little end. Apart from a couple of special engines, we run a plain bearing and that is where the trouble begins when the lubrication is not up to the job. These latter day lubes can be of good use to us, if they fill the bill in all the other areas, and if we make a little addition.

Consider all the criteria we have discussed from right back at the start of this fuel section. I have hinted at one oil that does all we require of it and I am sure most readers know to which oil I have been alluding. Did somebody mention castor oil? Give that man a new glow plug! Yes, castor oil; an extraction of the poisonous seeds of the euphorbiaceous Indian plant, *Ricinus communis*. This oil also contains a carboxylic acid—ricinoleic acid—that increases in ratio the more the castor bean is crushed. For this reason, the grade we should use is first pressings, low acid, medicinal or fine filtered industrial grade. A latter day saying, 'there's no such thing as a free lunch' applies to castor as, even though it fills all the criteria needed, it does have a couple of minor drawbacks. It comes out of the exhaust almost the same as it was before being mixed—gluggy and gooey—and sticks all over your model and, at engine idle, it creates a bit of internal carbon. These little problems are far outweighed when we consider all the good things it does for us, such as lubrication and cushioning the hammer effect. Okay, you are still concerned about the goo and carbon; let's strike a simple compromise: use a tested and proven synthetic oil as the main lubricant with the addition of some castor oil for added protection and cushioning. A mix that I have found very successful, and one that is also used by the leading Australian mixed fuel company, Magnum, is 10% high grade (proven) synthetic oil and 5% AA grade castor oil. (AA grade is first pressing, low acid.) The castor I use comes from the Castrol Oil company marketed under the name Castrol M and this is also first pressing, low acid, fine filtered castor oil (usual disclaimer). By using this 15% mix you are giving the engine the protection required, clean oil is still coming out the exhaust pipe but not making too much mess on your model. Note well! This is the absolute lowest oil proportion I would recommend for four-stroke engines unless, of course, a lower proportion is recommended by the manufacturer as in the case of (example) UK Magnum and OPS 120. It is your choice as to how much you use but, when you get up around the 20% castor proportion, you might find a little trouble with some engines kicking off the prop due to an hydraulic action, i.e. the compression is raised too much by the high oil content. I will give you a good, all round fuel mix recipe at the conclusion of this section but it is up to you if you wish to experiment with ratios and chemical content.

The OPS 120 features overhead cam operation.

Fuel content

The fuel content (the chemical that actually burns) of glow plug engines is methyl alcohol, commonly called methanol. There is a special reason for using this and why it is the only fuel that will do the job for our engines as they are designed. The ignition of glow plug engines is provided by an element inside the glow plug. The element is platinum or an alloy (metal mixture) of platinum and its derivatives such as rhodium and irridium. The plug is initially heated by an electric current which is disconnected when the engine starts. From then on the plug is heated partly by residual combustion heat and partly by a chemical action of the methanol contacting the platinum. This is a catalytic action that speeds up or intensifies the heating action. This chemical reaction is present in a few other chemicals but it is not sufficient to give controlled or proper running of the engine. At this point, no other chemical has been successful in replacing methanol as the main fuel for glow plug engines.

Other chemicals are sometimes added to fuel mixes to boost the performance but you pay dearly for this small gain. The most commonly used additive in glow plug fuels is nitro methane and, at current prices, you could purchase a nice .40 sport engine and a few sheets of balsa for the cost of a 4.5 litres (one gallon) can of it. A chemical aromatic, nitro methane is a form of fuel enhancer in as much as it actually increases the oxygen content of the existing fuel and this helps the combustion process. The proportion used in a four-stroke is generally 5 to 10% and its purpose is to help with the idle. With the non-firing stroke of a four-stroke, the glow plug tends to cool a little and idling can be unreliable. This problem is generally

English Laser with special tank and magneto set up for Old Timer events.

Shades of days past but still popular with a lot of modellers is spark ignition. This is an original coil and spark plug from early days.

more prevalent when the engine is new and a bit tight. It tends to be less of a problem when the engine has a few hours on it and I know of many modellers who have very sweet running four-strokers that have never known the taste of nitro. I have found that the addition of 5 to 10% petrol (any type) as part of the methanol percentage can assist the low idle running equally as well as nitro and this is a real plus. Nitro can wreak havoc with steel components such as bearings and gears when its residue is left to attack and cause rust whereas petrol is just the opposite—it tends to clean the lower end of the engine quite well. (Do not use petrol in YS 120 engines as it will destroy the silicone diaphragm in the pressure regulator. Check other exotic engines for similar components.) It is worth a bit of time experimenting with petrol as the cost is reasonable, it can do a good job and it leaves no harmful residues. When it is used the combustion temperature increases a few degrees and this assists the idle. It could also advance the ignition a little and you might need to experiment with plugs and the amount of petrol used to get the best combination. You need a low idle due to the coarse pitch props used on four-strokes when you consider, say, a 12×8 prop with the engine idling at 2,000rpm gives a (theoretical) speed of 15mph which is a fair rate, particularly for a floater type model.

Have we decided now what we are going to use as our fuel mix? Maybe F.A.I. standard which is 4:1 methanol/castor? Perhaps 15% castor/10% nitro methane/75% methanol? Sounds good so far. Well, I'm going to use my favourite of 10% Glo Glide synthetic/5% Castrol M/7% petrol and 78% methanol and I am going to ensure correct proportion by using metric measure which is so simple. If I want a small amount I make the percentages millilitres and I get 100ml of fuel. For one litre I multiply by each percentage by 10. Example: 100ml oil/50ml castor/70ml petrol/780ml methanol = 1,000ml which is one litre. So simple! If you want more than one litre, multiply each ml amount by the number of litres required, eg. 5 litres so, 5×100ml oil and so on.

The first start

Let's see; the engine is mounted, prop balanced, fuel selected so we're ready to start... almost. Did you check for foreign nasties in the fuel? It doesn't take much to block a fuel supply and that means pulling carby bits out—unnecessary work and wear—and searching for something that might be so small it is hard to see without an ocular aid. Save all the trouble—fit a fuel filter to the feed line from the tank to the carburettor. If you use this line for filling your tank, fill from *behind* the filter otherwise foreign nasties will lodge on the front of the filter element and be pushed/sucked into the carby as soon as the engine is choked or started. You might think this is overdoing it when I say that I have a filter on the bottom of the pickup in my fuel caddy (can for carrying fuel supply), one on the outlet of the tank tiller tube from the caddy and one on the supply tube from the tank to the engine. I can't recall ever having to strip one of my own carbies to find blockages though!

Now, are we ready? Hold on a moment—there's a large hole in your carby and it's right at the sucking section. Any chance it might suck in some grit? Sure is and it certainly does. The air intake of the carburettor is the most neglected part of a model engine and it can mean the difference between a rapid wear out and a long and happy life. A lot of modellers are in the habit of fitting fuel filters now but this only protects against a carby blockage. The path to the fuel jet in the spray bar is so

OS 120 Surpass equipped with a built-in fuel pump

When you have spent many hundreds of hours constructing a superb nine cylinder radial four-stroke, you are unlikely to risk damaging it by careless running-in.

small that anything that could pass down there with the fuel would be most unlikely to be large enough to do any damage to the engine. On the other hand, a rock the size of a housefly could wander down the throat of a carby and do some really nasty damage inside an engine. I have had quite a few engines for repair that have ingested grit and, in every case, the best the owner could hope for was a reasonably expensive repair bill for new parts.

One customer sent me his YS Yamada 120—a powerful, well made engine, with the complaint that it had lost compression. I sent him back a box of bits and a sad note, 'engine beyond repair'. The engine had been in a model that suffered a hard landing (crash) and course grit had found its way into the air intake of the carby. On starting the engine again, the grit had travelled around the disc valve scoring the lapped surfaces, entered the crankcase and chiselled several rows of burrs around the inside of the case (like a wood rasp), travelled up the induction tube from the crankcase and into the valve chamber where several pieces were pressed into the intake valve seat. One piece lodged under the exhaust valve and bent the sealing rim, some travelled with the piston, scoring it in the process, cut grooves around the ring and tore the chrome plating off the bore. Flakes of chrome entered the crankcase and wrecked the main bearings and then travelled the same path as the grit committing more horrors similar to the predecessor. The main bearings were at the point of collapse which caused the main shaft to spin in an erratic pattern and this wrecked the conrod. The carburettor and engine assembly bolts were undamaged. That is no exaggeration and the owner actually returned the engine to Yamada to show the extent of the damage that could have been prevented by the fitting of an air filter. An added benefit of an air filter is that it can reduce the engine noise by up to 2dB. I demonstrated the value of an air filter to a group of modellers and you can try the same experiment for yourself. Take the sponge element from a filter that has had a fair amount of use and rub it hard on a sheet of glass. You will generally find that the glass is etched with many fine scratches from the trapped grit. Consider what this same grit would have done to your engine if it had gone in with the fuel. Running the engine close to the ground as in a model is the worst situation. The propeller stirs up loads of dust and grit and the carby sucks it down. Fitting a filter is a small price to pay for a large saving.

Well, we have our glow driver or battery ready, fuel is in the tank and we are ready. You have got the tank at the correct height and the mixture set precisely at 4.5 parts of air to every one part of fuel haven't you? Not quite? Okay, let's check out these details so we don't wreck the engine on the first start.

Tank height for a four-stroke engine is more important than it is for a two-stroke as the 'stroker' does not have the suction or draw of a two-stroke. The level of the tank is always in reference to the centreline of the tank—not the outlet tube—and the reason for this is two-fold. One is because we are constantly changing the attitude of the aircraft, climb, dive, level and, maybe, some aerobatic manoeuvres. This combines with the other reason and that relates to head of pressure. The fuel weighs X amount (according to how much you have in a tank) and the tank is Y high. The height and weight of the fuel combines to make the head of pressure which will vary as the level drops. By keeping the jet of the carby around the alignment of centre of the tank, the pressure will be reasonably constant regardless of whether the model is right side up or inverted. Also, with the tank set up with the centreline in line with the jet (spraybar) the weight is reduced as the fuel has to be pushed/sucked uphill for half its height to get to the centre or the outlet. Whenever we use this method we will always have a problem with the mixture altering slightly as the pressure in the tank reduces, and it is for this reason that it is advisable to use muffler pressure to the tank and tune a little on the rich side with a full tank.

The Perry oscillating pump is ideal for four-stroke engines as it relies on only a little vibration to operate.

There are two really good methods of overcoming the lean run on low tank syndrome. Use a tank that is as wide and low as possible so that there is not so much fuel weight on the inlet (clunk) and, before the first flight of the day, fill the tank about ¼ to ⅓ and tune the engine on this tank level. When you fill up the engine will be just a trifle rich (which won't do any harm) but, when the tank is getting really low you know the engine won't run lean. For the clunk in the tank I find that the best, trouble free method is to fit a short length of silicone (15mm) to the pickup side of the engine feed tube then a length of copper or brass tube and, on the end of this, another short length (12mm) of silicone to connect the clunk weight. Keep the

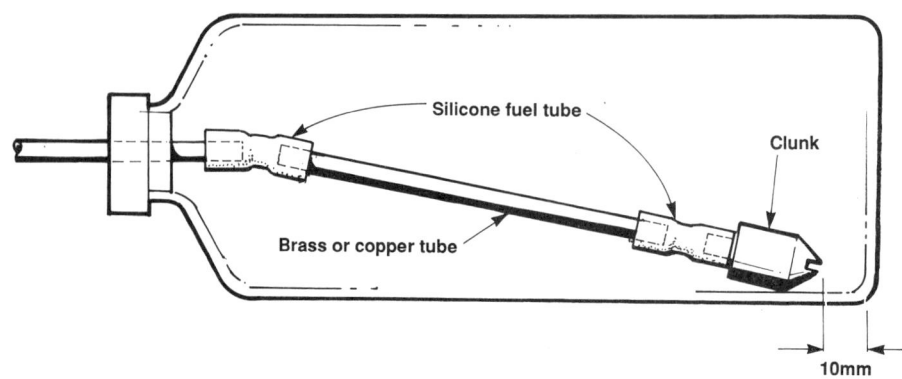

Clunk set-up for fuel tank.

clunk weight about 9mm from the rear wall of the tank so that it cannot suck up against the wall and starve the engine. This type of clunk works perfectly in all attitudes and won't kink or knot itself. So much for the tank; now, what about this 4.5 : 1 bit and how do we achieve this?

The correct air to fuel ratio for an engine running on methanol is 4.5 parts of air to one part of fuel. These two ingredients meet around the centre of the venturi (hole through the centre of the carburettor, sometimes smaller in the centre section and belled out at both ends) and combine to make a moist or misty gas that eventually enters the combustion chamber and explodes on ignition which we know as the engine firing. If the gas has too much fuel in the mix it is a rich mixture and conversely, a mix that does not have enough fuel is a lean mix. It is always better to have a very slightly rich mix at a prone position (that is, the model at rest or level) as the mixture leans slightly as the engine works under load such as a steep climb and the fuel has to be drawn uphill from the tank. The correct mixture ratio is obtained by adjusting the main needle valve which is a fine tapered needle entering into a small diameter tube. Beyond the point of the needle is the 'jet'—a hole of some shape (depending on the manufacturing design) where the fuel exits to join the air flow. A simple name for these components is needle and spraybar although this is more commonly used on engines that are not equipped with a variable speed carburettor. It is not possible to 'tune by mail' any engine as there are a number of variables such as the taper of the needle and the pitch (number of threads per inch) of the thread on the spraybar or barrel (needle holder). As a very rough guide it has been my experience that most engines run with about 1½ turns

open on the needle but I would never start a new engine on that setting. As I mentioned earlier, a lot of engines come from the factory with a nominal running setting on the carburettor but this might have been disturbed by a 'looker' in the model shop. Never take it that the settings are correct for the first start as many an engine has been ruined by the initial burst of life when it started, revved right up and died due to a lean setting. This lean burst on the first start can write off hours of life of the engine and, in extreme cases, write it right off from the start with a broken rod or seized piston.

Consult the instructions for a new engine as to correct first start procedures or needle settings. In the absence of such information, open the needle about 4 turns, start the engine and slowly wind the needle in (more precise details later). Before you start fiddling with the carby, familiarize yourself with the controls and adjustments and the function of each part. In the main you will have one of two styles—air bleed control or twin needle. An air bleed carby has only one needle and that is used for the high speed mixture. Somewhere on the front of the body you will see a small hole and, nearby, a bolt and spring. The bolt end actually enters the side of the hole to adjust the intake

The business end of a twin needle carburettor. This is the idle mixture needle.

A pair of air bleed carburettors showing the air hole and the adjustment screw above.

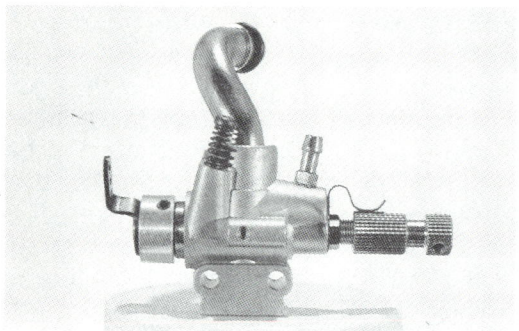

Modern style twin needle carburettor. Screw and spring is throttle barrel stop screw used for adjusting the idle speed-not mixture.

of air for idling when the carby barrel is almost closed. The idle speed is set by another bolt and spring on the top of the carby body and this acts as a stop for the barrel. To adjust the air bleed, bring the barrel to the stop for the engine to idle. If the rpm increases a little and the engine stops, the mixture is too lean—too much air, not enough fuel—so the needle needs to be screwed in just a little. If the engine rpm slows down and the engine stops, the mixture is too rich—too much fuel, not enough air—so screw the needle out a little. It only takes a little adjustment to make a difference so keep your adjustments to ¼ turn each time. Once the mixture is set you might find the rpm a little high or low for a reliable idle so readjust the throttle stop screw. A twin needle carby has an idle needle set in the end where the throttle arm is

attached and the same applies here as far as adjustment is concerned, ¼ turn each time at the most. One small difference is the direction of turning. The needle is screwed *out* for a richer mixture, *in* for a leaner mixture. The stop screw adjustment is the same as for the air bleed carby. When choking the engine—closing the air intake to rapidly suck up fuel—you definitely need to do it at full throttle with an air bleed carby otherwise you will only suck air through the front hole. A twin needle carby will suck at any position but it will suck faster at full throttle.

The other parts of the carby to familiarize yourself with are the throttle arm—it moves the barrel open and closed and is adjustable for position and the choke (when fitted)—for sucking the initial charge of fuel into a cold engine. The choke must not be left closed when attempting to start the engine and should not be used when an electric starter is used. If you overchoke the engine you stand a chance of creating an hydraulic lock—liquid fuel in the compression chamber—that will stop the engine turning. To try and force the engine over might result in serious damage. Even a mild overchoke is likely to cause the engine to backfire and throw the prop. If you encounter a lock problem, rock the prop clockwise several times until it can be turned over freely. Flick it several times in the correct direction to clear the chamber completely before attempting to start the engine.

Right then, now we are on friendly terms with the carby we can confidently go straight to the main needle and open it 4 turns. With the choke closed or a finger over the air intake of the carby, hold the prop and turn it over until you see the fuel come up the feed line and *just* reach the carby fuel nipple. Now, choke twice only by turning the prop two full turns. Release the choke. Set the carby to the idle position. Flick the prop twice and then connect the glow plug power supply. Hold the prop firmly and turn it over. You should feel a bump within two turns. As soon as you feel the bump, flick the prop smartly for an instant start. If you do not get any response from the engine after 3 or 4 flicks, disconnect the glow power and re-choke once only. Try again. If the engine has not started by now you will need to check the glow plug and that fuel is getting into the cylinder head (plug will be moist). Rectify any problem.

The engine is running at low throttle and we are going to advance the throttle. The engine should be running lumpy due to being rich so we start turning in the main needle about ¼ turn at a time and listen to the results for a few seconds. When the engine is running smoothly, open the needle again about ½ a turn so the engine continues to run but does so richly. Now follow the instructions for the correct running in procedure. In the absence of this, the following is a good method of running in and the time is considerably reduced. The benefit of a short run in period on the bench is that you are close at hand should the engine become distressed in any way and can take immediate remedial action. You are also familiarizing yourself with the controls and the running characteristics of the engine. Once you have the engine running rich but steady at full throttle and you are standing behind the prop (safety), hold the head with your thumb and forefinger. When it becomes uncomfortable to hold, stop the engine and let it get quite cold. Start again and repeat the procedure about ten times.

When you have completed this running in process your engine will be ready to fit to a model for flying but still at a slightly rich setting for two or three tanks of fuel. While you have it on the bench you can test the plug and the props. Fit the smallest prop you consider you might use and then start. When it is quite warm, take it down to idle and see if it will hold 2,000 to 2,500rpm. If the engine stops, restart and go down to idle again but leave the plug lead connected. If the engine will now idle smoothly it indicates that you need a hotter plug. If the engine still stops with the plug connected you need to tune the idle mixture. When you have the engine fitted in your model you can tune it up with the prop you are going to use and the fuel mixture you intend to run it on. Tune it up until the engine is running smoothly and then wind the needle out about ¼ turn so that is just a trifle rich. Remember, the exhaust sound is different to that of a two-stroke and it may sound a bit flat at first, but you will quickly learn when it sounds happy. A tachometer (rev counter) helps a lot here as you should enrich the mixture to the point where the engine drops about 200 to 300rpm. You should then condition yourself to leave the needles alone whenever you use the same prop, plug and fuel. The slight difference in running on very hot or cold days will hardly be noticed. Oh, alright then, you can adjust the needle for weather conditions everytime you adjust the needle in the carburettor of your (full-size) car. Does that make you happy now?

Mounting the beast

The first consideration you should make before you start laying down sticks for a new model is what method you are going to use for mounting the engine. For many reasons, the method that appeals to me most is the one-piece wooden crutch or two-piece beam. Both these end up much the same in the finished job, but the crutch has the advantage with strength and trueness of mounting surface. Consider that we are using a .60 or larger engine and the finished fuselage sides of the model will end up 6mm thick. We now need a 250mm length of 12mm thick hardwood that is the width of the fuselage minus 12mm (two side thicknesses). The wood must be machined on all surfaces and perfectly flat on one side (at least). With a pencil, mark a centreline along the length then mark the cutout for the engine area making the side beams 12mm wide. Make sure you radius the internal corners. From the end of the cutout mark the width of the firewall. From the rear firewall line mark the width of the front beams and taper these to the outer edge of the block at the other end. You should end up with a piece looking like a real crutch in miniature that is waiting for the legs to be pulled in at the bottom and the under arm rest to be fitted to the top. At this point the cutout will be far too wide for the beams of your engine but everything will be okay when we finish.

We now need a piece of Paxolin sheet around 4mm thick for a .60, up to 6mm for a 1.20, and this is available from

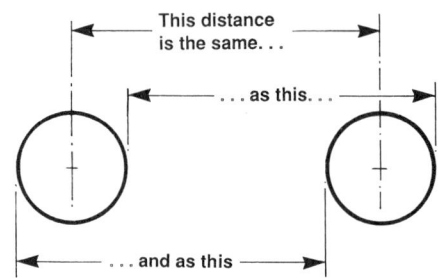

Engine mount marking pins.

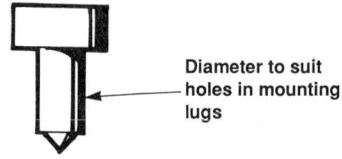

Diameter to suit holes in mounting lugs

Hardwood crutch for engine mount.

plastic and laminate suppliers or radio and electronic supply shops. Cut your piece to cover across the front beams from the front end of the beams to the front edge at the rear of the cutout. Using 4 bolts well spaced, mount the Paxolin to the beams using a scrap strip of Paxolin under each beam to act as a full length washer for the nuts on the bolts. (I prefer Nylock nuts on *all* mounting bolts in this area.) Now, mark out the Paxolin plate to suit your engine, cut out with a fine saw blade and sand

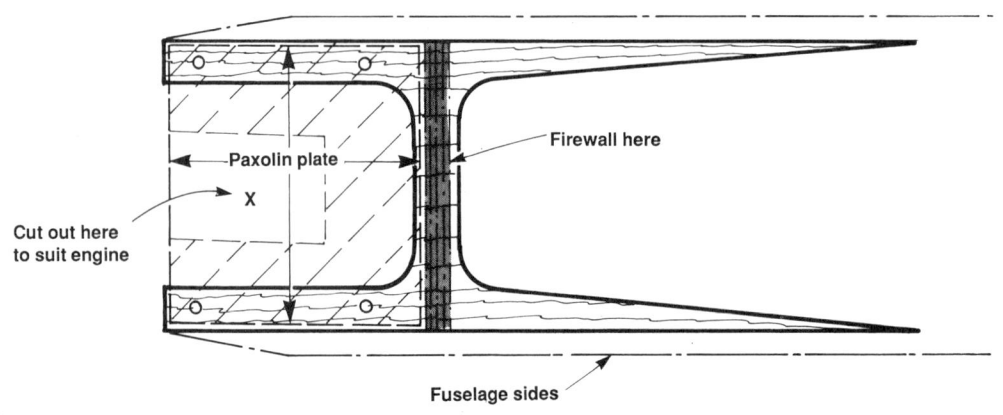

Measuring hole centres.

the edges. Fit the engine to the cutout and drill the mounting holes. This crutch can now be built into the front of your model adding great strength and rigidity and, in the event of a 'nasty' (crash), the Paxolin plate absorbs a lot of the crash energy and generally gives way before the engine suffers serious damage.

Direct mounting onto wooden beams is also a good method of mounting the engine, but the problem is that the beams have to enter into the fuselage beyond the firewall and tank placement can be a problem. Generally, the most preferred method of mounting by the majority of modellers is the ready-made radial mount. Not really a latter day idea as a lot of modellers might think, as this method was used way back when modellers made their own from wood and the unit (engine in mount) was secured to the model with rubber bands. It worked well then and could still work now on trainers for beginners to lessen the chance of engine damage in 'hard' landings. Radial mounts are available in cast or machined aluminium and fibre reinforced plastic. Some of the aluminium units look too sturdy to me as I consider the sides of the engine case could break out before the mount breaks in the case of a crash. The fibre/plastic units give good service, less likely to damage the engine and very light.

Before using a commercial mount or one you have made yourself, place a straightedge across the top of the mounting beams and hold up to the light. If you see wedges of light under the straightedge, the tops of the beams aren't true and will need machining or filing. Sit the engine in the mount and gently press down on each corner of the mounting beams to check for any rocking action. This will indicate that the beams are not true in some way. It is important that the top surfaces are precise as any variation will put pressure on the case of the engine and may cause distortion. With the working clearances of a model engine, any distortion of the case will cause premature wear or mechanical failure.

The underside of a lot of the commercial mounts are tapered which makes it a problem to use bolts and nuts to mount the engine, and it is about time the manufacturers addressed this problem. The USA idea is to use self-tapping screws to hold the engine and this is reasonable *if* the pilot hole is the correct size. Too big and the screw will eventually undo, too tight and you will screw the head off when tightening. I would not use this method as there is no locking device on the screw for added security. If you look around you will find a fibre/plastic mount that has parallel beams and an angle gusset on the side of each. These are ideal! If you have trouble obtaining these mounts, use the under tapered units, drill right through for the mounting holes and spotface or file a small flat for the nut and washer to seat on the tapered underside. Use Allen head cap screws for mounting your engine as they are easy to tighten with the Allen key and, being high tensile, can be tightened really tight without the worry of breaking or stripping. Another factor in their favour is that they don't stretch as a soft steel or brass bolt would, so there is less chance of them loosening.

Yes, I know, you want to know how to mark out the holes accurately. Well, I have a set of punches I made from silver steel (drill rod) and they are hardened. (Silver steel–heat to red, quench in cold water–reheat to straw colour and quench again.) The punches fit neatly in the mounting holes and are just a little longer than the thickness of the beam. A light tap on the top and the point marks the beam *exactly* in the centre of the hole. You can mark out the holes using a

With four-engined scale models it is prudent to fit larger capacity engines inboard and less powerful four-strokes outboard. So, this *Lancaster* might be powered by '60s' inboard and '40s' outboard.

jenny caliper and dividers and this is absolutely accurate, if you are also. (To measure the centre of two holes of the same diameter, measure from the edge of one hole to the corresponding edge of the other hole and that measurement is the centre distance between the holes.) Another simple method is to paint the top of the mount beams with correcting fluid (white fluid for painting over typing or writing mistakes obtainable from stationery shops) and let dry (almost instant). Clean the underside of the engine beams and sit on the mount in the correct position. Mark the position of the ends of the engine beams with a fine pencil line so that it can be relocated in the exact position. Remove the engine and press the underside of each beam onto a stamp pad, check that it has a thin coating of ink then replace it on the mount, holding it down firmly for about thirty seconds. The correcting fluid will absorb the ink in an accurate pattern of the mount with the holes clear. Centre the holes and drill accurately.

Safety check

When the engine is mounted in the model you will need to do a few flight checks after the first flight. Check the tightness of the mounting bolts all round and the propeller retaining nut. Do this again after the third flight and then before the start of each flying day. I know it might be a bit inconvenient to remove the spinner and cowl each time but consider how inconvenient it would be if the engine fell out of the model or the prop came loose and the engine had a shaft run.

Chapter 6
Problems and maintenance

MODEL ENGINES today are, in the main, troublefree but you could encounter problems with the nut that turns the propeller at times (that's *you*). If you should strike trouble, check possible causes listed below, but try them one at a time so that you know what *is* causing the problem.

Engine will not start

Needle valve closed or blockage in valve.	Open needle or clear blockage by blowing or pumping fuel through jet.
Plug not glowing.	Check plug and power supply.
Throttle barrel closed.	Adjust stop screw or trim on radio.
Valves not closing.	Check clearance and operation of valves.
No compression.	Check valves as above. Ring stuck in piston. Engine worn out.
No fuel.	Check fuel supply. Remove tube from carby and check that fuel flows.
Water in fuel.	Check engine with fuel from another modeller.

Engine starts but stops when battery is disconnected

Plug too cold.	Use hotter plug.
Plug no good.	Replace with new plug.
Water in fuel.	As above.

Hard to start or lack of power

Valve problems.	See above.
Prop too light.	Use heavier prop or fit metal spinner.
Crankshaft bearings worn or corroded.	Replace.
Crankcase vent blocked.	Clear blockage.

Air leak in intake manifold.	Rectify.
Excess carbon on exhaust valve stem or in exhaust chamber.	Decarbonise chemically.

Engine will not idle

Carby not adjusted correctly.	Adjust.
Plug too cold.	Hotter plug.
Cold fuel.	Replace 5–10% methanol with petrol or nitro methane.
Engine tight.	Run in more.
Prop too light.	See above.

Engine throws prop or backfires consistently

Lean mix.	Adjust main needle.
Choke closed.	Check.
Prop too light.	See above.
Plug too hot.	Use cooler plug.
Plug too far into combustion chamber.	Fit another washer.
Too much oil in fuel.	Read fuel section.
Engine overloaded.	Smaller prop.
Prop loose.	Tighten.
Acetone in fuel.	Use pure methanol.

(*Some methanol based racing fuels contain a small amount of acetone as standard*)

Compression too high.	Fit head gasket.
Engine flooded.	Rock prop backwards until clear.

Excessive vibration

Prop/spinner not balanced.	Balance.
Prop not tracking.	True up blades.
Main bearings worn or corroded.	Replace.
Bent crankshaft.	Straighten or replace.
Engine loose on mount or mount loose.	Tighten.
High compression.	Gasket under head.
Damage to engine due to crash/home modifications or clamping in vice.	Repair.
Oversize muffler unsupported.	Rectify.
Incorrect timing.	Check.

Other considerations

The timing of a four-stroke engine is set by the manufacturer and keyed or locked into a permanent position until the engine is dismantled. Make sure you know the correct timing procedure or observe and understand the timing marks or gears or toothed pulleys before you dismantle an engine. All timing marks relate to the piston being set on Top Dead Centre (TDC)—top of its travel. To attempt to alter the timing, 'just to see' what might happen, could result in the piston striking one or both of the valves with disastrous consequences.

Except where specified by the manufacturer, the breather nipple is not to be used for any other purpose. It is there to exhaust the pressure and excess oil from the crankcase and for no other reason. It cannot be used for pressurizing the fuel tank. If you fit a length of tubing to the nipple to exit out the cowl, make sure the oil is actually getting out. If the tubing is too long, the oil cannot reach the end before it is sucked in again and your crankcase will gradually fill with waste oil causing strain on the working parts.

Engine maintenance

If you keep an engine long enough it will certainly reach a point where it will require some maintenance or replacement of worn or broken parts. If you don't feel confident stripping the engine yourself, check with your hobbyshop dealer as he possibly has an engine repairer contact or you might check with your club. If you decide to do the job yourself then we will look at a few hints on easy procedures and preventative maintenance.

The first job we will look at is prevention of corrosion to steel parts such as ballraces, gears and crankshafts and, at the same time, prevention of gumming up the engine. A four-stroke does not have the flushing facilities a two-stroke has with the fresh charge of fuel going through the crankcase on each induction. As such, corrosive residues tend to accumulate in the crankcase of the engine and do nasty things. It is reasonably common knowledge that the main bearings in a four-stroke engine can be chewed out with rust rather quickly and I have repaired engines where all the bearings and

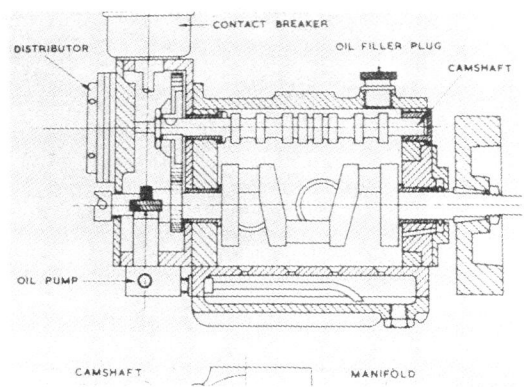

Section view on a blueprint for a three-bearing, double throw crankshaft.

The engine had only half an hour on the clock, was put aside for a couple of weeks and the fuel residues did the rest. An almost total internal rebuild could have been avoided by the use of some after run treatment.

gears had to be replaced and the crankshaft dressed in the lathe to remove rust corrosion. The common cry is 'NITRO'—nitro methane—as the residues of this chemical form nitric acid. I agree that this can be a difficult problem but it is not on its own. Castor has ricinoleic acid in its makeup and methanol is hygroscopic (absorbs moisture from the air), so you can end up with a mixture of acids and water and other by-products of combustion just waiting their chance to do dastardly deeds to the internal necessaries. This nastiness and the gumming up can be prevented by giving the engine a little attention when you finish flying for the day.

The way you do it is to inject some neutralising agent into the case that will clean out the gunk and leave a protective coating. There are several after run treatments on the market that may do the job but you don't know until you try them. I looked for something that was doing a similar job and had been proven by large usage and past reputation. I also wanted something that would not be diluted by methanol and would provide protection for an extended period and leave the engine ready for an immediate start regardless of how long it had been stored. I found the answer in (car) automatic trans-

The joys of use of your engine. This model flies and sounds quite realistic in the air.

mission fluid. Superb lubrication for fine tolerance machinery, prevents corrosion and does not dry out or go gummy. I mix 50/50 petrol and transmission fluid and this is injected into the engine via the breather nipple with the aid of a syringe. I pump in about 20cc or so and then suck it in and out about half a dozen times. While the fluid is in the engine, turn the prop over 5 or 6 times to get some on the liner, gudgeon and ring. It is thin enough to pass the ring and coat the top of the piston. After a good sloshing around suck the remainder out with the syringe and, where possible, store the model nose down so that any remaining liquid will run out the front bearing onto the thick wad of newspaper underneath. This gives the engine a thorough internal cleaning and leaves a thin coating of the oil over all parts. For the rocker gear on top of the engine I like to use a graphite oil or one of the super oil additives available for full-size cars— the additives you buy in a tin that turns your old banger into a sweet running dream. Most of these are superb, modern lubricants that have a long life and a tenacious grip on wearing surfaces. At the very least, a few drops of light oil about once a month when you check the tappet adjustment and don't forget to dribble a few drops down the pushrod tubes for the pushrod sockets in the cam followers. Most of the time a reasonable amount of oil is bypassed through the cam followers and it finds its way to the top end, but this is not always the case. A regular check of the area could save some expensive repairs. Check if the needle in the carby is fitted with an O ring and, if so, a touch of light

Tappet adjustment is regular maintenance for a four-stroke as gap must be correct for optimum running. Feeler gauge is inserted on the top of the valve stem to check the clearance.

Nestled in the nose of the *Tiggy* is an OS .20 four-stroke with onboard ignition for reliable idle while running inverted.

oil here once a month will make everything work freely and prevent the O ring being ruined by galling (tearing) on a dry surface.

When you do your monthly tappet check it is a good idea to check the bolts in the engine for tightness (engine cold) as a loose bolt anywhere on an engine could be the start of a big problem. Every bolt is important and has a specific job to do. If one is loose, the part it helps hold together could warp beyond use or cause other serious damage to the engine. Set the tappets according to the directions of the manufacturer or, where not available, set with just enough gap to insert a piece of writing paper but not enough for two thicknesses. With practice, the gap can be set without gauges as you need a just discernible free movement. As the engine heats up the gap increases just a little as it is subject to the 'steel doughnut' principle. (This is a good discussion point at the field sometime: if a steel ring (doughnut) is heated to red heat, will the hole in the middle be larger or smaller than it was when cold?) It is important that you check the tappet adjustment regularly as both too much and too little gap will rob the engine of power. No gap might cause the valve rim to burn and too much will cause the ends of the valve to be hammered to the point where they cannot be removed from the head as the burr on the stem will not fit through the valve guides.

De-carbonizing

After some time the engine might have a build up of carbon in the exhaust valve

This delightful little Tiger Moth is very light for a span of 48 inches. Flight is just like the real thing.

Pistons and other 'combustion area' components eventually become carbonised. There are safe chemical cleaners on the market and it is also possible to use synthetic lubricants, such as 'UK Aerosynth' to reduce the effects of carbonisation.

chamber and the combustion chamber. It needs to be removed chemically. Do not use scrapers or other tools for this purpose as you are sure to cause some damage to the metal. Remove the head from the engine and remove both valves. It is a good idea to place the head in a large, clear plastic bag when removing the valves in case a valve collet or spring decides to take a flight into places never known. (Yes, your hands should also be inside the bag as it is mighty difficult to do the job otherwise.) Check the valves for seating indicated by a narrow, shiny ring on the angled rim. Do not damage this surface or the seat in the head in any way as it is necessary for an airtight seal of the valve in the head during combustion. Coat the valves, combustion chamber and exhaust valve chamber with gel type paint stripper (having previously ascertained that it is safe on aluminium, most are), or aluminium cookware cleaner. *Do not use oven cleaner unless you know it is safe on aluminium as some contain caustic soda and this is deadly on aluminium and its alloys.* Leave the cleaner for the recommended time—about ½ hour—and clean off with water, remembering that the cleaner burns flesh and can blind you if you get a splash of it in the eye. If the job is not

properly done, repeat the operation. If the carbon on the exhaust valve stem resists attempts to chemically clean it you can resort to the edge of a lolly stick or similar wooden implement, remembering that too much force will bend the valve stem rendering it useless. When the carbon is off, scrub the parts with a Brillo (typical) pad and detergent, wash well and boil in clean water for 3 or 4 minutes. Then tip into a strainer to drain for about 10 seconds then onto a wad of clean cloth where the heat will dry it (generally). Oil the valve stems and reassemble again in the plastic bag. Set the completed head aside protected from dust while you remove the back plate from the engine and wash it out with petrol. When clean, inspect the bearings for corrosion or damage. If all is well, and the engine had reasonable compression before stripping it down, it is not necessary or advisable to strip it any further so give the bottom end a good dose of thin oil or auto transmission fluid and close up. Replace the head and tighten the bolts in the following pattern. Visually number the headbolts 1 to 6 in a clockwise fashion. Just nip up 1 then 4 then 6 then 3 then 5 then 2 then back to 1 to tighten and repeat the sequence. For a 5 bolt head the sequence is 1–3–5–2–4–1 and for a

The OS .80 was a modernized, bored out 60 with enclosed rockers. Note the rear camshaft housing.

The Surpass series from OS was the second wave of four-strokes and featured increased power with easier starting.

One and only effort from Kalt was the Kalt .45. Runs well.

4 bolt head or front/backplate the sequence is 1–3–4–2–1. By following the tightening sequence applicable to your job you are eliminating the problem of uneven tightening and warping the part being tightened.

Replacing main bearings

The prop driver and crankshaft generally have to be removed by using heat, a press or a suitable puller. The case is then heated and the bearings are removed. A good safe method of heating the case is with a heat shrink gun as is used for covering material. The paint stripper heat guns are also suitable for this job. To fit the new bearings, the case is again heated and the bearings aligned with the crankshaft and slid into place. Do not attempt to press bearings into a cold case unless you are absolutely familiar with this operation. I have kept this section to an explanation only of the procedure as it is one that, if not done correctly, could cause considerable damage to expensive engine parts. For the removal of the piston and liner and the replacement of the piston ring, I advise you to seek the assistance of a modeller experienced in the procedure to show you the correct steps.

The final word

Your model four-stroke engine is a fine example of precision, miniature engineering and deserves a little extra care. It is an investment in your enjoyment of your leisure time and it will continue to please you for a long time if you take care as set out in this book. Remember the Chinese idiom 'softly, softly, catchee monkey' as the exhaust note of the engine may be deceptively quiet but the bite of the prop is realistically hard. It may sound 'softly, softly' but it will 'catchee' finger if you are careless!

Always tune a little on the rich side, don't go too far down on the oil content and balance the prop.

Almost forgot! The hole in the 'hot metal doughnut' *increases* in size (bet that causes some arguments!). Proof? Consider a shrink fit. Why does the ballrace fall out when you heat the crankcase? Think about it.

Appendix
Manufacturers' and suppliers' addresses

UK

AGC Sales, London Road, Apsley, Hemel Hempstead, Hertfordshire HP3 9ST (Laser engines plus spares)

Argus Specialist Publications/Argus Books/Argus Plans, Argus House, Boundary Way, Hemel Hempstead, Hertfordshire HP2 7ST.
Tel: (0442) 66551

Irvine Engines Ltd., Unit 2, Brunswick Industrial Estate, Brunswick Way, New Southgate, London N11 1JL. Tel: (081) 361 1123 (O.S. engines plus spares)

MacGregor Industries Ltd., Canal Estate, Langley, Berkshire SL3 6ED. Tel: (0753) 42251 (Saito engine plus spares)

Modeland, 219 New North Road, Hainault, Essex. Tel: (081) 500 3891

Len Rawle, Practical Scale, 132 Berry Lane, Chorleywood, Hertfordshire WD3 4BT (Large capacity engines Zenoah, Titan, Tartan)

Ripmax Models Ltd., Ripmax Corner, Green Street, Enfield, Middlesex, EN3 7SL. Tel: (081) 804 8272 (ENYA engines and spares)

USA and other

Altech Marketing Inc. P.O. Box 286, Fords, New Jersey 08863, USA (ENYA engines)

Great Planes, PO Box 4021, Champaign, Illinois 61824-4021, USA (O.S. engines)

F Kavan, Lindenstrasse 56, D-8500 Nurnberg 10, West Germany

O.S. Engines Manufacturing Co. Ltd., 6-15 3-chome, Imagawa, Higashisumi-yashi-ku, Osaka 546, Japan

Subscribe now...

here's 3 good reasons why!

Within each issue these three informative magazines provide the expertise, and inspiration you need to keep abreast of developments in the exciting field of model aviation.

With regular new designs to build, practical features that take the mysteries out of construction, reports and detailed descriptions of the techniques and ideas of the pioneering aircraft modellers all over the world – they respresent three of the very best reasons for taking out a subscription. You need never miss a single issue or a single minute of aeromodelling pleasure again!

SUBSCRIPTION RATES

	U.K.	Europe	Middle East	Far East	Rest of World
RCM&E *Published monthly*	£18.00	£28.20	£28.90	£33.80	£31.90
Radio Modeller *Published monthly*	£18.00	£24.10	£24.50	£27.45	£26.30
Aeromodeller *Published monthly*	£23.40	£28.20	£28.40	£30.20	£28.70

Airmail Rates on Request

Your remittance with delivery details should be sent to:

The Subscriptions Manager **(CG/64)**
Argus Specialist Publications
Argus House
Boundary Way
Hemel Hempstead
Herts
HP2 7ST